Sustainable Surface Engineering Approaches for Industrial Applications

The text describes the fundamental concepts of usability and user experience, and how they are defined, designed for, and eventually measured and evaluated. It covers the surface treatment of bioimplant materials by the electrical discharge machining process and the influence of surface texture on the tribological behavior of artificial joints. It showcases the experimental investigation of the surface hardness of two dissimilar metal-welded joints.

This book:

- Discusses advanced surface treatment methods such as diamond-like carbon coatings and nanocoatings
- Explains important topics in the areas of laser surface engineering, surface parameters of wire arc welding, and bioimplant surface treatments
- Presents industrial applications in the areas of automobiles, such as cylinders, biomedical devices, and steel surfaces
- Explores surface modification of engineering materials using heat treatment and coating
- Highlights the effect of heat treatment on the tribological and mechanical properties of polymeric materials

This book is primarily written for senior undergraduate, graduate students, and academic researchers in fields including mechanical engineering, industrial and production engineering, manufacturing engineering, and aerospace engineering.

Application Tribology

Materials, Design, Manufacturing and Testing

The proposed book series will comprehensively discuss the basics and fundamentals of friction, wear and lubrication to the emerging and special topics, such as nano-tribology, bio-tribology, food tribology, tribology of fabrics, tribology of extreme environments, tribo-corrosion, wear modelling, monitoring and testing of wear. Aimed at senior undergraduate, graduate students, academic researchers and professionals, the proposed series will focus on presenting the latest research in the field of tribology. It will also explore different aspects such as materials, design, manufacturing, and testing.

Lubricant Technology: Fundamentals and Applications
Edited by V Vasu, Upendra Maurya and Gourav Khajuria

Sustainable Surface Engineering Approaches for Industrial Applications: Surface Treatments and Surface Textures
Shubrajit Bhaumik, Viorel Paleu, V. Jayakumar and Emrullah Hakan Kaleli

Sustainable Surface Engineering Approaches for Industrial Applications: Sustainable Coating Solutions
Shubrajit Bhaumik, Viorel Paleu, V. Jayakumar and Emrullah Hakan Kaleli

Sustainable Surface Engineering Approaches for Industrial Applications

Surface Treatments and Surface Textures

Edited by
Shubrajit Bhaumik, Viorel Paleu, V. Jayakumar, and Emrullah Hakan Kaleli

CRC Press
Taylor & Francis Group
Boca Raton London New York

CRC Press is an imprint of the
Taylor & Francis Group, an **informa** business

Designed cover image: Shutterstock

First edition published 2026
by CRC Press
2385 NW Executive Center Drive, Suite 320, Boca Raton FL 33431

and by CRC Press
4 Park Square, Milton Park, Abingdon, Oxon, OX14 4RN

CRC Press is an imprint of Taylor & Francis Group, LLC

ISBN: 978-1-032-59002-8 (hbk)
ISBN: 978-1-041-19774-4 (pbk)
ISBN: 978-1-003-71336-4 (ebk)

DOI: 10.1201/9781003713364

Typeset in Sabon
by Deanta Global Publishing Services, Chennai, India

Contents

About the editors

Shubrajit Bhaumik is an Associate rofessor in the Department of Mechanical Engineering, Amrita School of Engineering, Amrita Vishwa Vidyapeetham, Chennai Campus. Bhaumik established and is the center head of the Tribology and Interactive Surfaces Research Laboratory (TRISUL), which is dedicated to industrial application-oriented research. Bhaumik has been working with several industries in different areas of tribology since 2006 and has completed several projects that have resulted in energy savings for industries. Bhaumik has hands-on experience in hard-face welding, open-gear inspections, and lubricants. His research interests include nano and green tribology, energy-efficient lubricants, and solutions for minimizing friction and wear. Bhaumik is experienced with several wear test rigs, particularly nanolubricants and energy-efficient lubricants. He has successfully implemented energy-efficient nanolubricant concepts in real-time applications, which has benefited several industries. Bhaumik works in allied fields of tribology such as lubricants, polymers, and fretting wear. Bhaumik has published in several reputed journals of tribology, engineering, and book chapters. Recently, an Indian patent has been granted to him in the area of tribology and a few are in the pipeline. A tribology research team led by Bhaumik developed an AI-integrated bearing failure detection app that has been featured in the reputed dailies *The Hindu* and *The Times of India*. These apps are becoming famous in industries. He has won several awards and accolades in the area of tribology including Best Research Award—Malaysian International Tribology Conference 2015, a gold medal, and DST SERB Government of India fellowship. He is also the international ambassador of Surface Venture, United Kingdom, a not-for-profit organization that aims to spread the knowledge of tribology among industry and the academic community. Bhaumik is also an editorial board member of several tribology journals.

Viorel Paleu has been a habilitated full professor of mechanical engineering since October 2021 at the "Gheorghe Asachi" Technical University of Iasi, Romania, where he began his academic career in 1996 and earned his Ph.D. with magna cum laude distinction in 2002. Over two decades, he advanced through all academic ranks at the university's Faculty of Mechanical Engineering. He conducted doctoral and post-doctoral research at INSA Lyon, France, collaborating with industrial partners and securing multiple international research grants. He has led five national research projects and contributed to over 20 national and international initiatives. His research spans areas such as tribology, lubrication, rolling bearings, coatings, machine elements, materials, and diagnostics, with publications in prominent scientific journals. He has served on over 800 peer-review panels of international journals and as a committee member or chair for major international conferences. Currently, Paleu serves as vice dean for research and quality at the Faculty of Mechanical Engineering. He is also vice president of the Romanian Tribology Association (ART) and has been president of the Romanian Transmissions Association (ROAMET) since 2008.

V. Jayakumar is principal, School of Engineering and Computing, Chennai; and professor of mechanical engineering, Amrita Vishwa Vidyapeetham, Chennai Campus. With over 25 years of teaching experience, including 7 years abroad, he has served as visiting faculty in Oman, Eritrea, and Ethiopia. He holds a B.E. in mechanical engineering, M.E. in industrial engineering, MBA, and a postgraduate diploma in operations research. He earned his Ph.D. from the College of Engineering, Guindy, Anna University. Jayakumar has authored 10 textbooks, 4 of which are prescribed in Indian universities, and published over 85 research papers in Scopus-indexed journals. His research focuses on cellular manufacturing systems, friction stir welding, composite materials, advanced manufacturing systems, and production and operations management.

Emrullah Hakan Kaleli is a professor at Yildiz Technical University, Mechanical Engineering Department of Automotive Division, in Istanbul. His main fields are internal combustion engines (gasoline and diesel engines), tribology, wear, lubrication, and friction. He also researches engine tribology, nanotribology, lubrication, friction wear, engine oils, additives, coatings, drain interval and additive effects on the sliding surfaces, surface analysis, nanoadditives, and nanocoatings as he has many special international publications on these subjects. He has been a member of scientific committees and advisory committees of many international conferences.

Contributors

Arivazhagan, B.
Materials Development and Technology Group
Metallurgy and Materials Group
Indira Gandhi Centre for Atomic Research
Kalpakkam, India

Aydin, Mehmet Esat
Mechanical Engineering Department
Konya Technical University Institute of Graduate Studies
Konya, Turkey

Bagci, Mehmet
Mechanical Engineering Department
Engineering and Natural Science Faculty
Konya Technical University
Konya, Turkey

Bayraktar, Esathan
Metallurgical and Materials Engineering Department
Konya Technical University Institute of Graduate Studies
Konya, Turkey

Bhaumik, Shubrajit
Department of Mechanical Engineering
Amrita School of Engineering
Amrita Vishwa Vidyapeetham
Chennai, India

Chattopadhyaya, Somnath
Department of Mechanical Engineering
Indian Institute of Technology (Indian School of Mines)
Dhanbad, Jharkhand

Dharmalingam G
Vel Tech Rangarajan Dr. Sagunthala R&D Institute of Science and Technology
Avadi, Chennai, India

Govindaraju, M.
Department of Mechanical Engineering
Amrita School of Engineering
Amrita Vishwa Vidyapeetham
Coimbatore, India

Jegannathan, P.
Department of Mechanical Engineering
Amrita School of Engineering
Amrita Vishwa Vidyapeetham
Coimbatore, India

Kaleli, Emrullah Hakan
Faculty of Mechanical Engineering
Yildiz Technical University
Istanbul, Turkey

Kannan, Rajesh K.
Department of Mechanical Engineering
Amrita School of Engineering
Amrita Vishwa Vidyapeetham
Coimbatore, India

Kori, Jaya
Department of Mechanical Engineering
Indian Institute of Technology (Indian School of Mines)
Dhanbad, Jharkhand

Kumar, Sai Prasanna J.V.
Vel Tech Rangarajan Dr. Sagunthala R&D Institute of Science and Technology
Avadi, Chennai, India

Kundu, Soumyajit
CSIR-Central Mechanical Engineering Research Institute (CSIR-CMERI)
Durgapur, India

Manivasagam, Thirugnasambandam G.
Centerof Excellence in Advanced Materials & Green Technologies, Amrita School of Engineering
Coimbatore, Amrita Vishwa Vidyapeetham, India

Mukherjee, Diya
CSIR-Central Mechanical Engineering Research Institute (CSIR-CMERI)
Durgapur, India

Mukherjee, Manidipto
CSIR-Central Mechanical Engineering Research Institute (CSIR-CMERI)
Durgapur, India

Muralikannan, R.
Department of Mechanical Engineering
Sethu Institute of Technology
Kariapatti, Tamil Nadu, India

Nampoothiri, Krishnadas Narayanan
Department of Mechanical Engineering
Amrita School of Engineering
Amrita Vishwa Vidyapeetham
Chennai, Tamil Nadu, India

Pramanik, Sumit
Department of Mechanical Engineering
College of Engineering and Technology
SRM Institute of Science and Technology
Tamil Nadu, India

Rathinavel, S.
Department of Aeronautical Engineering
Nehru Institute of Engineering and Technology
Coimbatore, Tamil Nadu, India

SenthilKumar, S.
Department of Mechanical Engineering
SRM Institute of Science and Technology
Tiruchirappalli, Tamil Nadu, India

Senthilkumar, T.S.
Centre for Material Science
Department of Mechanical Engineering
Karpagam Academy of Higher Education
Coimbatore, India

Shree, Maanu R.
Department of Mechanical Engineering
Amrita School of Engineering
Amrita Vishwa Vidyapeetham
Coimbatore, India

Tiwari, Yoshit
CSIR-Central Mechanical Engineering Research Institute (CSIR-CMERI)
Durgapur, India

Tripathi, Rupam
Department of Mechanical Engineering
Indian Institute of Technology (Indian School of Mines)
Dhanbad, Jharkhand

Unuvar, Erdem
YUMAK Automotive Products Industry and Trade Limited Company
Konya, Turkey

Uysal Veli
Faculty of Mechanical Engineering
Yildiz Technical University
Istanbul, Turkey

Vignesh, Vaira R.
Department of Mechanical Engineering
Amrita School of Engineering
Amrita Vishwa Vidyapeetham
Coimbatore, India

Vikram, Kamlendra
Department of Mechanical Engineering
College of Engineering and Technology
SRM Institute of Science and Technology
Chennai, Tamil Nadu, India

Chapter 1

Experimental investigation of the effects of engine cylinder honing as tribological aspect on performance, wear, and lubrication

Emrullah Hakan Kaleli and Veli Uysal

1.1 INTRODUCTION

It is well known that the presence of the honing scratches/grooves in cylinder liner surfaces is intended and desired as they improve the lubrication and retain the debris, reducing the piston assembly/liner friction and consequently improving the fuel economy and longevity of the internal combustion (IC) engines [1].

Automotive industry components, such as cylinder sleeves, crankpins and camshaft lobes, valve seats and valve guides, valve body, and high-pressure pump components, need lubrication, which demands a unique surface texture. This is achieved by plateau honing, obtained in two machining operations: prehoning or rough honing with large abrasive grain size and subsequent finishing or fine honing with small abrasive grain size. Rough honing induces deep creases of specific geometry for effective lubricant retention. The smooth surface between the deep creases is machined to reduce wear and enable sealing–finishing honing [2].

The honing process is usually applied to the machined workpieces to improve surface quality and dimensional accuracy as a final machining process. The honing operation occurs with the abrasive effect of the erosive grains into small particles. Corundum, silicon carbide, boron nitride, and diamond are used to give abrasive properties to honing stones [3]. The abrasive particles are joined using materials such as ceramic crayons, synthetic resins, fungi, and carbon nitride. The honing process is a machining process that is applied to workpieces with honing stones, and workpieces contact at low pressures such as 10–100 N/cm² [4].

In other words, honing is an *abrasive machining* process that produces a *precision surface* on a *metal* workpiece by scrubbing an *abrasive* stone against it along a controlled path. It is primarily used to improve the geometric form of a surface, but may also improve the *surface texture* [5]. Honed surfaces show good tribological properties, high contact ratios, and low roughness in connection to the defined crossing canals, which can absorb lubricants, improve the running-in characteristics, and increase the

DOI: 10.1201/9781003713364-1

life of tribologically stressed parts. In addition to the high surface quality, the technology of long-stroke honing makes it possible to produce internal cylindrical surfaces with high dimensional and form accuracy. Fields of application include the finish machining of slide bearings, cylinder liners, and piston cylinders [6].

During the machining process, the honing stones are in contact with the workpiece by means of mechanically or hydraulically driven systems. The honing heads consist of a body and 3 to 12 pieces of honing stones, which are located equally spaced around the body, and nonabrasive guides. Guides that are not abrasive parts of the honing head are manufactured from bronze, cast iron, and sintered materials [4].

The honing process can be divided into several specific classes with regard to applying type. The main ones are longitudinal stroke honing, short stroke honing, center plunge honing, profile honing, surface honing, and gear honing techniques. The longitudinal stroke honing process is applicable to connecting rod big and small ends, cylinder walls, brake bearings, and roller bearing guide holes. The longitudinal stroke honing process has advantages in regard to short processing time, high chip removal capacity, and high quality of cylindricity and dimensional accuracy. However, there are some difficulties with center-to-center alignment of the honing head and workpiece [3]. The research of this work covers the cylinder walls honing angle, which is created during the longitudinal stroke honing process.

Two basic movements occur in the honing process. One of them is the axial movement, which causes the oscillation, and the other is the rotational movement, which enables the rotation of the honing stones circularly. During the honing process, the hydraulically or mechanically driving system doesn't have pressure at the honing stones. In other words, the honing stones are not displaced during this time. Thus, in the honing process, firstly, the honing head gets within the workpiece, and the honing head's diameter is smaller than the workpiece hole; secondly, the honing stones are driven hydraulically or mechanically and they expand, then the honing stones and workpiece remain in contact. Thus, the honing stones that come into contact with the workpiece generate removal chips as a result of the rotational and translational movements. The cutting parameters must be precisely adjusted, and measurement defects must be minimized, which arise from residual stresses and thermal loads, because the honing process is the finish surface machining process. For this purpose, the operational surface temperature shouldn't exceed 100°C [4]. Therefore, the cutting fluid must be precisely adjusted. The cutting fluid also has tasks such as removing the chips from the surface and preventing the tool from smearing.

If the contact pressure increases during the honing process, the cutting tool's ability to penetrate increases. In this case, the honing stone is exposed to greater stresses. As the contact pressure increases, the honing stone will be exposed to greater stresses, which will increase wear, increase the surface roughness, and increase the amount of removal chip quantity. The curvature of the cylinder walls will begin to improve slightly, but then start to worsen with a stronger tendency [3]. For this reason, there is an upper limit to increasing the contact pressure.

In the longitudinal stroke honing process, the honing head makes both rotational and translational movements with the shaft to which it is attached. At this time, as the honing head moves downward, the corners of the honing stone create scratches on the surface of the workpiece along the direction of rotation and the direction of oscillation. Then the same issue occurs in the transverse direction during the upward stroke. Ultimately, the honing marks generate a honing cross-hatch angle, which characteristically forms in the honing process. The honing cross-hatch angle is an angle that is generated by two transverse honing traces. Slits formed during the honing process serve as an oil reservoir, working as a microbearing [7]. The honing cross-hatch angle directly affects the tribological properties between the cylinder wall and the ring [8]. The honing cross-hatch angle can be changed by altering the oscillation speed of the honing head and the speed in the rotational direction of the honing head. If the oscillation speed decreases or the rotational speed increases, the honing angle is moved to a smaller value.

If conventional methods are used during the longitudinal stroke honing process, a generally constant honing cross-hatch angle will be generated at the workpiece. However, during the change in the direction of motion, a slightly smaller honing angle is obtained as a natural result of the linear acceleration of the honing head. This is due to the fact that the feed rate velocity drops to zero and then accelerates in the opposite direction while the oscillation direction is changing [3].

Taking into account previous formal testing conditions of colorful protective layers of additives under boundary lubrication conditions and changing honing angles, the cylinder surface honing is optimized, especially to reduce friction and wear, and increase engine life.

Tribometer tests were performed on a steel ball and honed cylinder liner using a fully formulated oil in the boundary lubrication condition. It was found that a cross-hatch 20° honing angle was effective in lowering the coefficient of friction and increasing surface protection and lubrication by forming a protective layer on the rubbing surfaces.

1.2 EXPERIMENTAL DETAILS

Tribological performance of the cross-hatch honed liners were compared with 5W-40 synthetic engine oil (Table 1.1) according to formal, well-established test parameters, utilizing a customized reciprocating tribometer setup, as showed in Figure 1.1a–c. The tribometer is driven by a servomotor, and 800 data values per second were collected from the load cell, using a data logger, directly connected to a monitor. Each stroke lasted 0.8 s (640 data per second), and in this way, the average coefficient of friction (COF)

Table 1.1 Properties of 5W-40 synthetic engine oil

Total base number (TBN)	11.28 mg KOH/g
Flash point	217°C
Viscosity @ 100°C	13.59 cSt
Viscosity @ 40°C	84.27 cSt

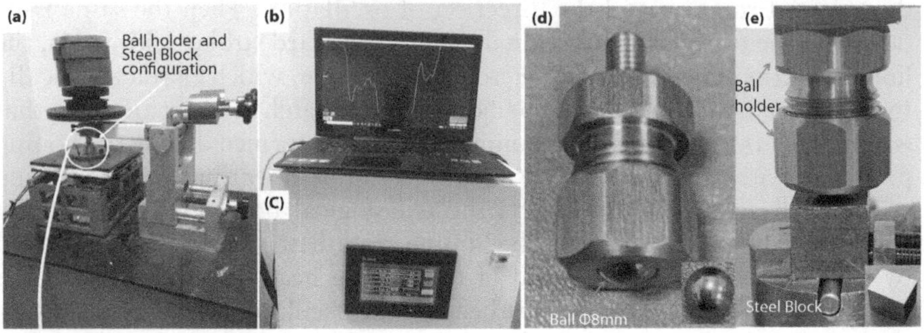

Figure 1.1 Reciprocating tribotest setup: (a) a general view, (b) recording of accurate friction force data measurement, (c) heating and temperature control and monitoring, (d) ball placed in the holder, (e) ball holder and liner integrated configuration.

of the wear track during the measurement was estimated. All tests were carried out under boundary lubrication conditions at 100°C. Normal load during the tribological measurements was set to be 60.5 N. Sliding speed or reciprocating velocity and stroke were 0.055 m/s and 8 mm, respectively. The temperature of the liner was set up to 100°C, using an advanced heater equipped with a digital controller mounted on the setup. To run a 21-minute tribotest measurement, three drops of engine oil were dropped on the surface of each honed liner. Then, the 100Cr6 steel ball was rubbed on it under boundary lubrication conditions, as displayed in Figure 1.1d,e. Data acquisition was done by specific software developed using MATLAB, which was used in order to filter any vibrational noise and to calculate the average COF by plotting the final COF curve as a function of time.

1.2.1 Preparation of cylinder liner specimens for honing and tribotesting

The Anadolu Motor Antor 6 LD 400 diesel engine and its technical specifications to be used for experimental work and its technical specifications are presented in Figure 1.2a. It is an air-cooled, single-cylinder engine. The cylinder barrel test specimens were cut accurately by the dry cutting method. The cutting process was performed at Dünya Kalıp in Istanbul, Turkey, using CNC milling process technology *without damaging the honed surfaces.* The substrate temperature was considered and the temperature controlled

(a)

Engine	Specification
Model diesel	6 LD 400
Cylinder volüme	395 cm^3
Cylinder diameter	86 mm
Compression ratio	18.1
Max. speed	3600 rpm
Max. power	8.5 HP
Max. torque	2 kgm @2200 d/dk
Fuel capacity	4.5 lt
Fuel consumption	220 gr/HP/hour
Oil consumption	13 gr/hour
Oil capacity	1.2 lt
Dry weight	45 kg

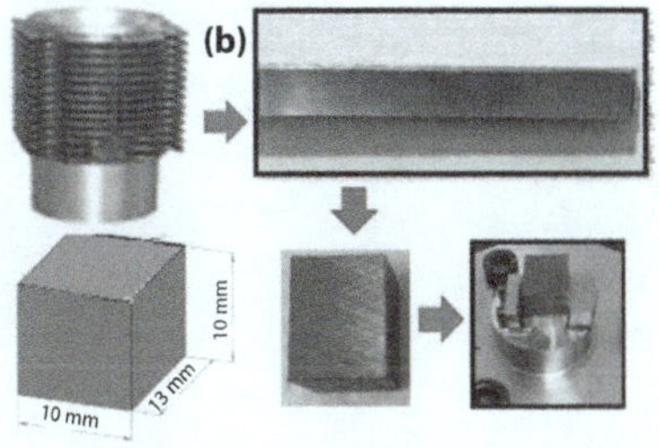

Figure 1.2 Anadolu Motor Antor 6 LD 400 diesel engine and (a) its technical specifications to be used for tribometer tests. (b) Cutting of cylinder barrel of diesel engine and mounting in tribometer.

during the cutting process due to possible material property changes at cylinder–liner substrates induced by heat. During the cutting process, the surface temperature was controlled by an infrared digital thermometer, and measurements showed that surface temperature had not exceeded 49°C. The cylinder–liner substrate dimensions were 10×13×10 mm (w×l×h) as presented in Figure 1.2b.

Table 1.2 shows chemical composition of the nontested ball, which is a low-alloy martensitic chrome steel, AISI 52100 steel ball (100 Cr6). The liner material of this engine is gray cast iron. Chemical analysis (analyzed on 01.12.2016) of cylinder material provided by the Anadolu Engine Group for LD 400 engine is given in Table 1.3.

Figure 1.3a,b presents the cylinder liner bore of the Antor 6 LD 400 engine to be honed with five different angles (20°, 30°, 40°, 45°, and 60°). According to information from the Anadolu Engine Group, the original (regular) honing angle of Antor LD400 is 45°. As a nonhoned cylinder sleeve exists in the Anadolu Engine Group, spare parts were purchased, and they were honed for 20°, 30°, 40°, and 60° honing angles at Anadolu Motor in Istanbul. An example of the 20° honing angle is shown in Figure 1.3c,d.

Figure 1.4 presents the measurement of the five different cross-hatch honing angles (20°, 30°, 40°, 45°, 60°) using a special digital microscope. They were tested in a reciprocating tribometer for tribological purposes.

Surface analysis of the balls and liner sliding pairs was carried out as well through a fully digital optical microscope in order to compare the degree of deformation occurring upon testing the honed liner samples, while 2D/3D roughness parameters were measured by a 3D high-resolution optical profilometer, before and after tests. All roughness measurements were carried out with an X10 Olympus objective choosing a high-quality section in software. The liner surface is plateau honed. Ra and Sa values were considered for comparison. 2D/3D roughness parameters can be seen clearly in figures as they are less than 1 μm (item no. 135 Honing 6 patched the honing is 0.05 to 1.6 μm), which conforms well with UKAS (the national accreditation body for the United Kingdom).

Table 1.2 Chemical composition of the new ball 100 Cr6 nontested

Sample name	*%C*	*%Si*	*%Mn*	*%P*	*%S*	*%Cr*	*%Ni*	*%Mo*	*%Cu*
Low-alloy martensitic chrome steel, AISI 52100 (100 Cr6)	0.95–1.05	0.15–0.35	0.25–0.45	0.025 max	0.025 max	1.40–1.65	0.30 max	0.08 max	0.20 max

Table 1.3 Chemical analysis of cylinder material (gray cast iron)

Element	%	*Element*	%	*Element*	%
Carbon C	3.2000	Silicon (Si)	2.59000	Iron (Fe)	92.44082
Phosphorous (P)	0.33319	Sulfur (S)	0.02737	Manganese (Mn)	0.70000
Molibdenum (Mo)	0.02700	Nickel (Ni)	0.01900	Chromium (Cr)	0.43700
Cobalt (Co)	0.00343	Copper (Cu)	0.11600	Aluminum (Al)	0.00010
Titanium (Ti)	0.08497	Vanadium (V)	0.01401	Niobium (Nb)	0.00297
Tin (Sn)	0.00359	Magnesium (Mg)	0.00045	Wolfram (W)	0.00010

Figure 1.3 (a, b) The cylinder liner of Antor 6 LD 400 engine to be honed. (c, d) Example of 20° honing angle.

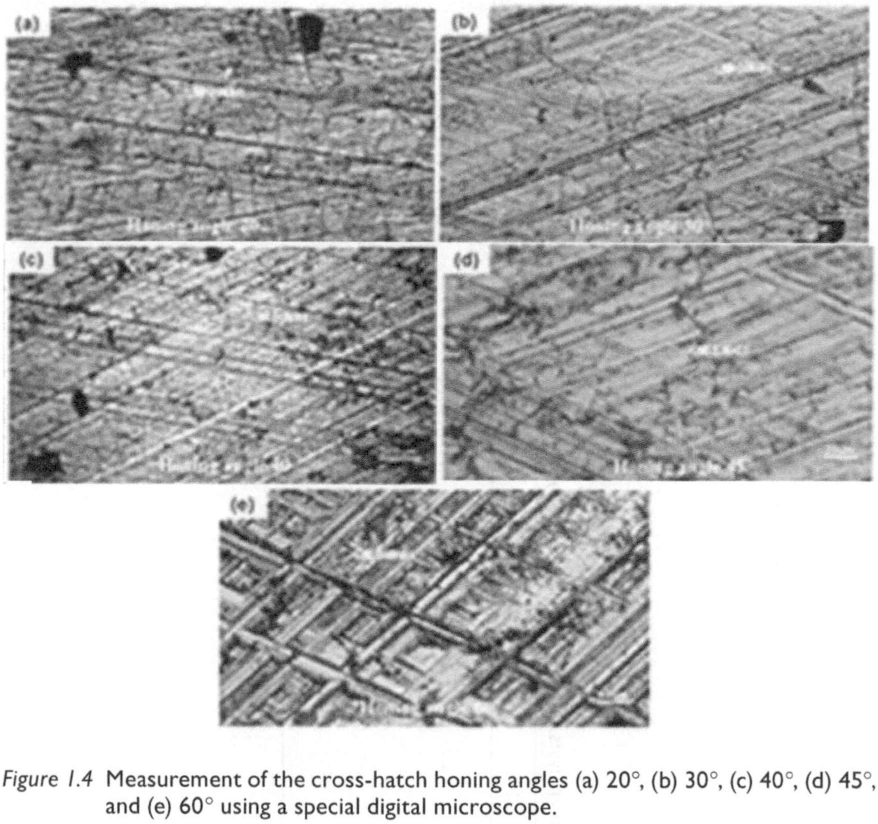

Figure 1.4 Measurement of the cross-hatch honing angles (a) 20°, (b) 30°, (c) 40°, (d) 45°, and (e) 60° using a special digital microscope.

Figure 1.5 shows the 2D and 3D roughness (Ra: 0.024 μm, Sa: 0.386 μm), SEM image, and EDS spectra of the nontested ball. The numbers on the EDS in the white frames correspond to the EDS spectrum number. Elemental analysis is listed in the Table 1.4.

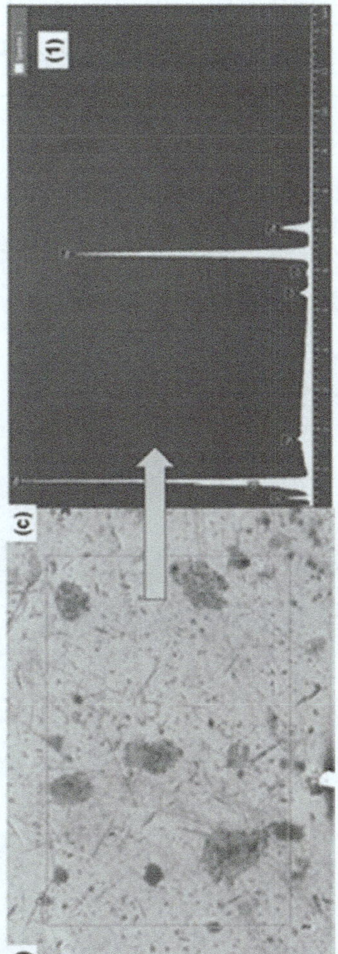

Figure 1.5 (a) 2D and 3D roughness, (b) SEM image, and (c) EDS spectra of the new ball to be tested.

Table 1.4 Elemental analysis of the new ball no tested by EDS

Spectrum number	*Sample name*	*C*	*O*	*Al*	*Si*	*P*	*Ca*	*Mn*	*Cr*	*Fe*	*Total*
1	New ball (nontested)	7	1	0	0	0	0	0	2	90	100

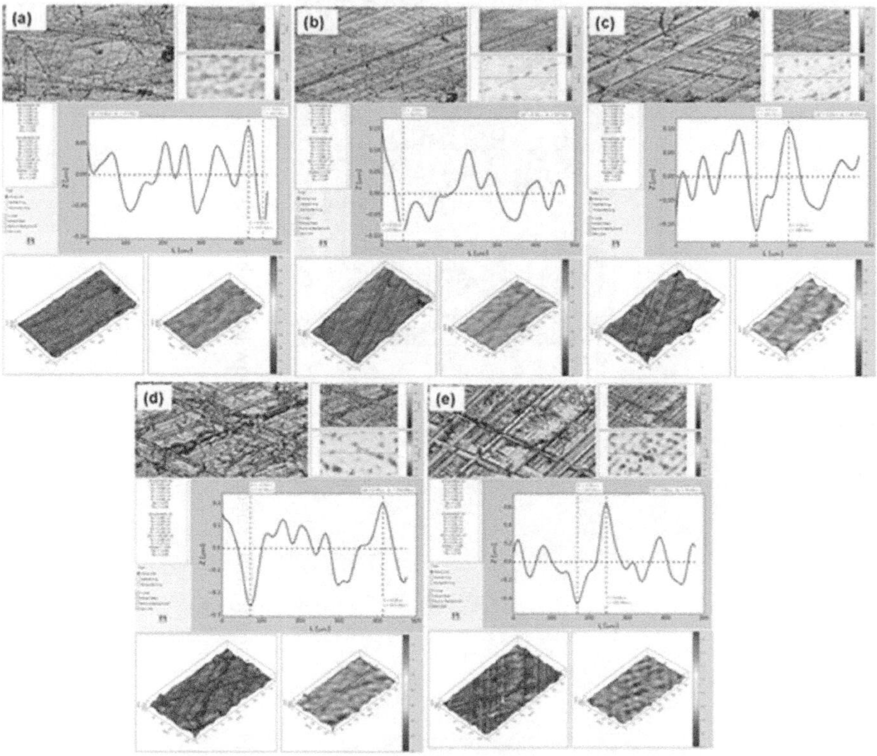

Figure 1.6 2D and 3D roughness of the nontested and honed with five different angles: (a) 20°, (b) 30°, (c) 40°, (d) 45°, and (e) 60°.

Figure 1.6 shows the 2D and 3D roughness of the nontested and honed with five different angles: 20°, 30°, 40°, 45°, and 60° cylinder liners. 2D and 3D roughness values are presented in Table 1.5.

The nontested 20° honed liner presented the lowest value according to the nontested 30°, 40°, 45°, and 60° honed cylinder liners.

Table 1.5 2D and 3D roughness values of the nontested 20°, 30°, 40°, 45°, 60° honed liner specimens

Honing angle	*Sa (nm)*	*Ra (nm)*
20°	**0.301**	**0.030**
30°	0.446	0.038
40°	0.598	0.044
45°	0.651	0.089
60°	0.653	0.169

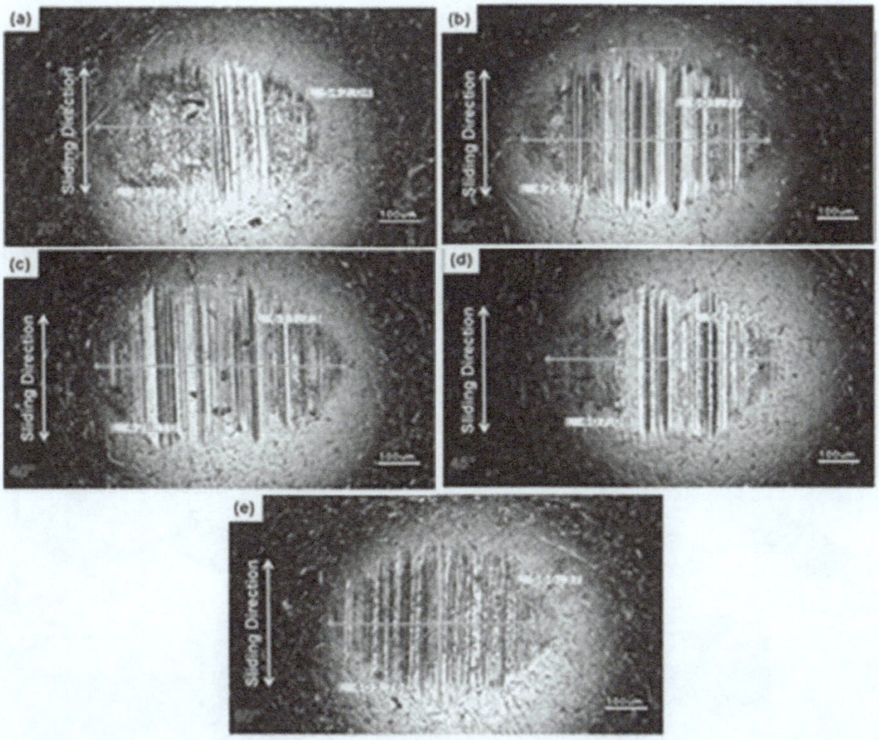

Figure 1.7 Optical digital images of the wear scar of the balls tested on the (a) 20°, (b) 30°, (c) 40°, (d) 45°, and (e) 60° honed cylinder liners.

1.3 RESULTS AND DISCUSSION

Figure 1.7 presents the optical digital images of the wear scar of the balls rubbed on the five different honing angles.

As seen in Table 1.6, the wear scar diameter of the ball presented the lowest value horizontally and vertically at a 20° angle compared to 30°, 40°, 45°, and 60° honed cylinder liners.

Table 1.6 Wear scar diameter of the balls tested on the 20°, 30°, 40°, 45°, and 60° honed cylinder liner

Honing angle	*Wear scar diameter, horizontally (μm)*	*Wear scar diameter, vertically (μm)*
20°	**542.04**	**361.73**
30°	601.77	426.99
40°	611.73	428.10
45°	563.05	407.08
60°	566.37	403.76

Figure 1.8 presents 2D–3D roughness digital optical microscopy of the tested ball on 20°, 30°, 40°, 45°, and 60° honed liners. Roughness parameters such as the Sa and Ra values of the wear scar presented the lowest value in 20° honed liner according to the tested 30°, 40°, 45°, and 60° honed cylinder liners as presented in Table 1.7.

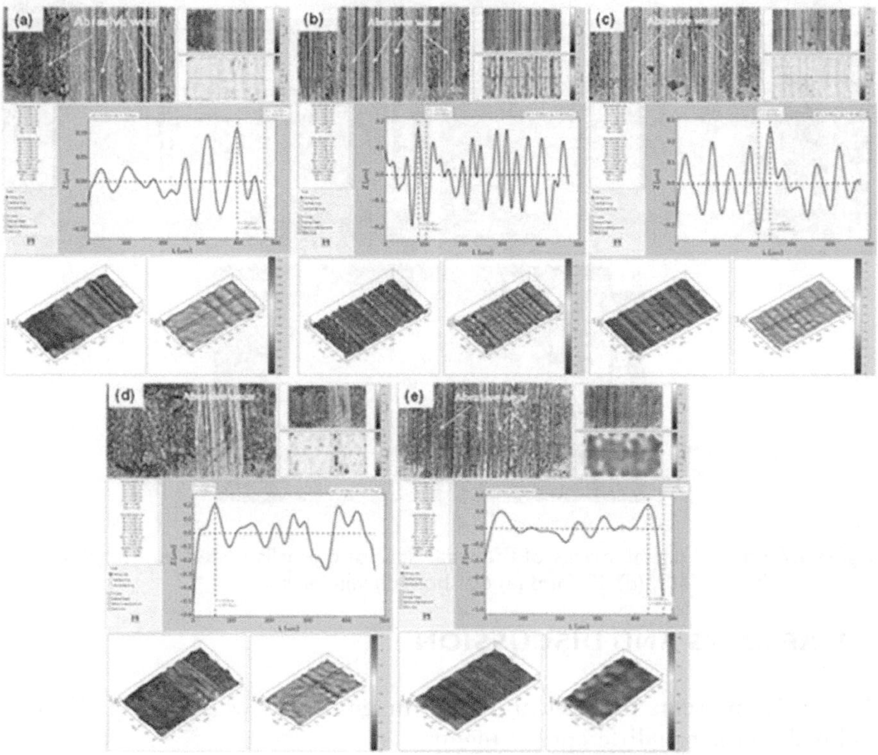

Figure 1.8 2D and 3D roughness values of the wear scar of the balls tested on the 20°, 30°, 40°, 45°, and 60° honed cylinder liners.

Table 1.7 2D and 3D roughness values of the wear scar of the balls tested on 20°, 30°, 40°, 45° and 60° honed cylinder liners

Honing angle	*Sa* (μm)	*Ra* (μm)
New ball (nontested)	0.386	0.024
20°	**0.351**	**0.030**
30°	0.406	0.072
40°	0.730	0.084
45°	0.923	0.095
60°	1.303	0.097

Figure 1.9 presents the comparison of the optical digital images of the wear track of the 20°, 30°, 40°, 45°, and 60° honed liner specimens rubbed under a steel ball. As seen in Table 1.8, the wear scar distance of the 20° honed angle presented the lowest value related to 30°, 40°, 45°, and 60° honed cylinder liners.

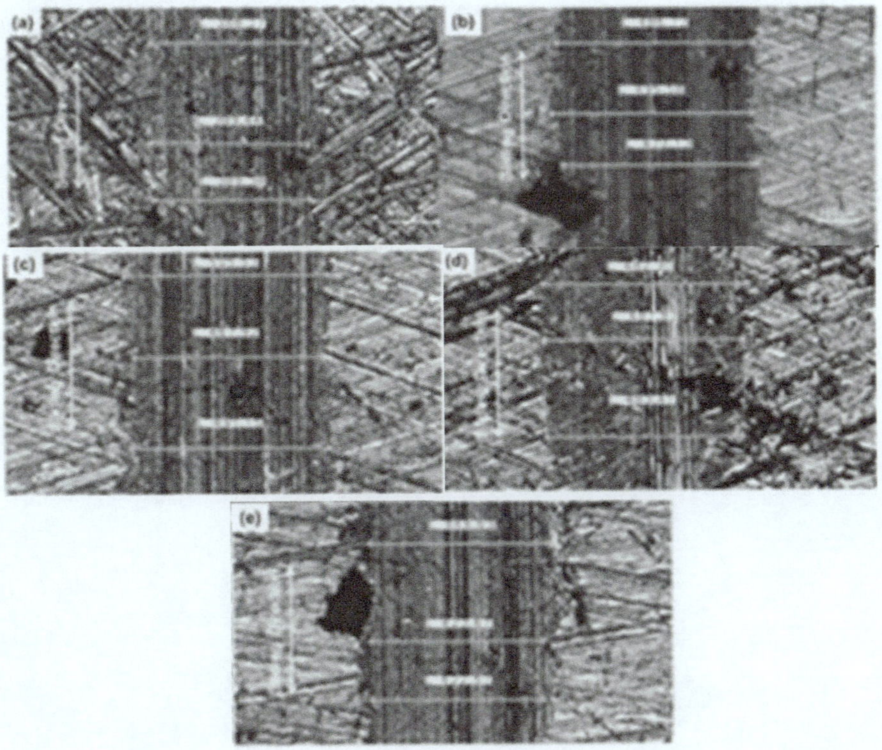

Figure 1.9 Optical digital images of the worn surface of the tested (a) 20°, (b) 30°, (c) 40°, (d) 45°, and (e) 60° honed cylinder liners under a steel ball.

Table 1.8 Maximum and minimum wear scar distance of the honed 20°, 30°, 40°, 45°, and 60° honed liner specimens tested under steel balls

Honing angle	*Maximum wear scar distance (μm)*	*Minimum wear scar distance Min (μm)*
20°	**389.44**	**388.27**
30°	495.58	470.13
40°	505.53	467.92
45°	487.83	482.30
60°	444.69	422.57

Figure 1.10 presents 2D–3D roughness digital optical microscopy of the 20°, 30°, 40°, 45°, and 60° honed tested liners. Roughness parameters such as Sa and Ra values of the wear scar distance presented the lowest value in 20° honed liner according to the tested 30°, 40°, 45°, and 60° honed cylinder liners as presented in Table 1.9.

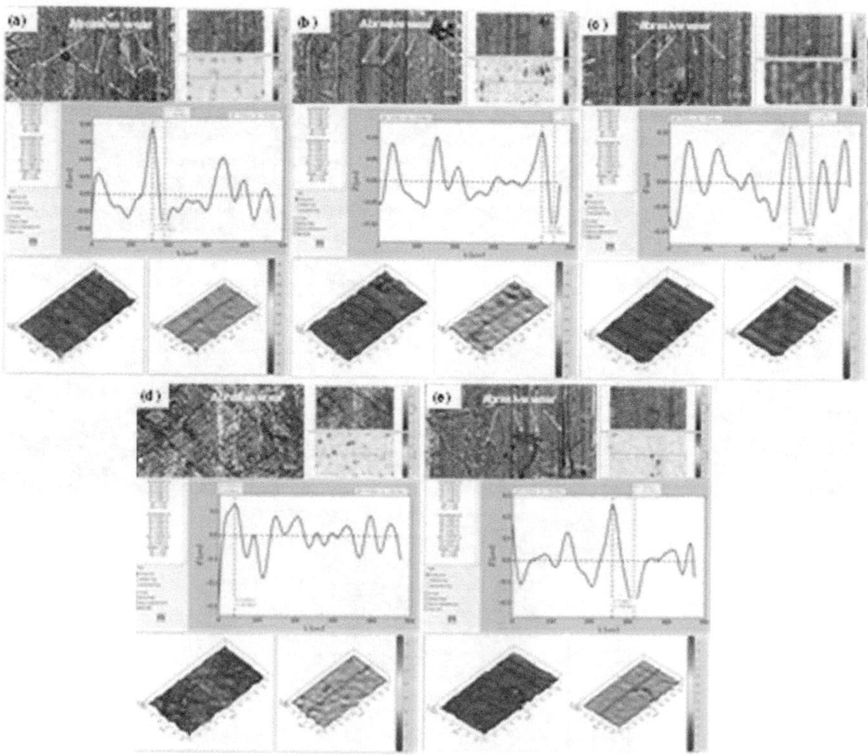

Figure 1.10 2D–3D roughness digital optical microscopy of the 20°, 30°, 40°, 45°, and 60° honed tested liners.

Table 1.9 2D and 3D roughness values of the nontested 20°, 30°, 40°, 45°, and 60° honed liner specimens tested under a steel ball

Honing angle	*Sa (μm)*	*Ra (μm)*
20°	**0.233**	**0.018**
30°	0.276	0.036
40°	0.341	0.043
45°	0.421	0.051
60°	0.723	0.065

As the test with the 20° honed cylinder liner demonstrated better results, SEM images and EDS spectra have been measured in this specimen. Figure 1.11 shows SEM images and EDS spectra of the rubbed surface of the ball tested on the 20° honed cylinder liner and the 20° honed cylinder liner tested under a steel ball. Elemental analysis is listed in Table 1.10 with spectrum numbers 21, 22, 23, 80, 81, and 82. Additive elements, such as Ca, Zn, and P, from engine oil well protected the rubbed surfaces.

Taking into account SEM–EDS results, elemental analysis presented accumulation of additives such as Zn, Ca, and P content formed in the protective layers and reduced coefficient of friction. That means the 20° honed cylinder liner provided more effective lubrication between the sliding parts. Therefore, it is apparent that it plays a key role toward the diminution of friction and the wear track reduction, contributing to surface protection against shear stress through the delamination effect [9].

1.3.1 Coefficient of friction results

Friction tests were repeated three times. The horizontal lines represent the average COF value, which is automatically plotted by the MATLAB-assisted specific software of the reciprocating tribotest setup. The average COF value exhibited a reduction between 5.93% and 12.42% (0.09780 lowest, 0.11673 highest) as estimated upon the completion of the experimental reciprocating tribotests in triplicate.

Figure 1.12 compares the coefficient of friction data obtained from the first, second, and third experimental reciprocating tribotests of the 20°, 30°, 40°, 45°, and 60° honed tested liners.

The lowest COF value is obtained with the test of 20° honed liner.

Figure 1.13 is the comparison of the average coefficient of friction data of the three repeated tribotests from the 20°, 30°, 40°, 45°, and 60° honed tested liners. The average lowest COF value is obtained with the test of 20° honed liner.

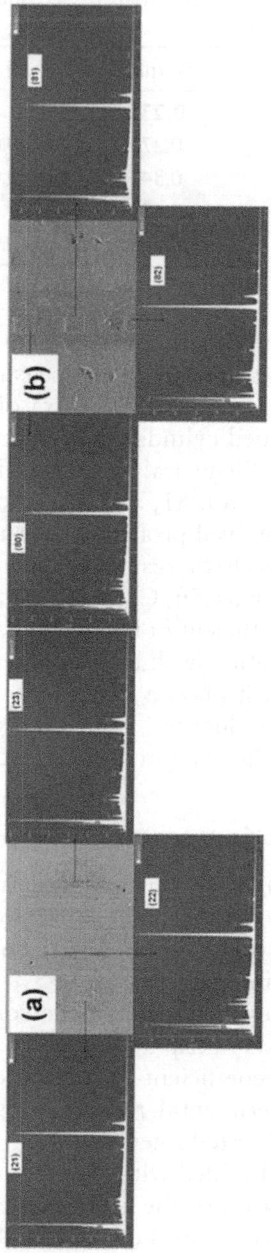

Figure 1.11 SEM image and EDS spectra of (a) the ball tested on the 20° honed cylinder liner and (b) the 20° honed cylinder liner tested under a steel ball.

Table 1.10 An overview of the elemental analysis of the wear scar of the rubbing parts by EDS

Spectrum number	*Sample name*	*C*	*O*	*Al*	*Si*	*P*	*Ca*	*Mn*	*Fe*	*Zn*	*Total*
21	Wear scar of the	4.66	4.21	0.07	0.35	0.74	0.15	0.48	87.31	2.03	100
22	ball rubbed on	3.9	9.2	0.08	0.38	2.36	0.92	0.52	76.25	6.39	100
23	20° honed liner	4.4	6.22	0.07	0.3	1.49	0.66	0.54	81.63	4.69	100
80	Wear scar of 20°	5.62	7.86	0	1.85	0.71	0.31	0.79	80.71	2.14	100
81	honed liner	4.93	6.36	0	1.9	0.69	0.33	0.81	82.72	2.26	100
82		5.24	8.69	0	1.73	1.05	0.39	0.74	79.36	2.8	100

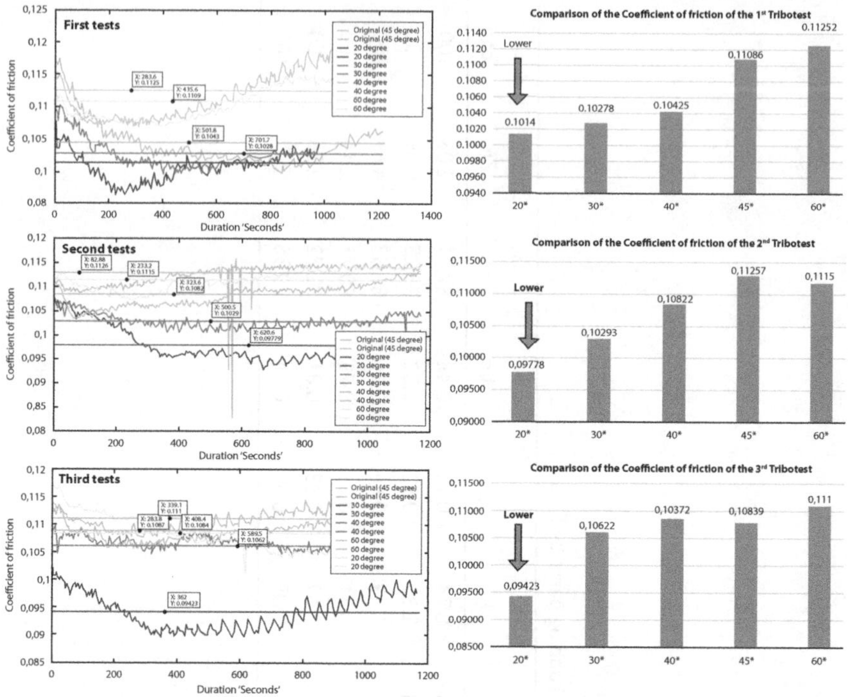

Figure 1.12 COF values obtained after the first, second, and third tribotests compared to coefficient of friction data obtained from experimental reciprocating tribotests of the 20°, 30°, 40°, 45°, and 60° honed tested liners as a function of testing time.

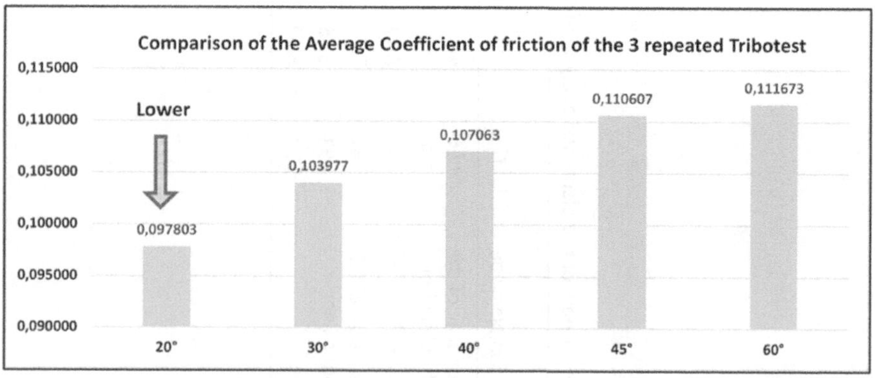

Figure 1.13 Comparison of the average coefficient of friction data obtained from three repeated experimental reciprocating tribotests of the 20°, 30°, 40°, 45°, and 60° honed liners.

It is generally possible to mention three different processes in the honing of cylinder liners of internal combustion engines. They are rough honing, finishing honing, and plateau honing. The valley and peak roughness values of the workpiece surface decrease during rough honing and fine honing operations. Fine honing is the same as rough honing in principle and has lower surface roughness values depending on the properties of the used honing stones. The plateau honing process is the last process in honing cylinder liners of internal combustion engines. With the plateau honing process, the peak roughness values of the cylinder can be seriously reduced while the valley roughness value of the cylinder does not change much. As a practical approach after this process, a relatively flat surface profile with low roughness oscillations is formed while the roughness of the valley forms microlevel bearings. In this way, valley roughness volumes can assume the role of lubricating bearings. On the other hand, depending on the reduction of the peak roughness, the effective contact area of the piston and cylinder will increase, and the surface pressure will decrease. The wear characteristic will also get better accordingly.

There are several methods for evaluating the surface roughness values numerically. In general, it is not possible for us to have foresight about the negative and positive deviations of roughness made by the arithmetic average, which does not allow for a healthy comparison. For this reason, examining the arithmetic mean or the valley bottom–peak roughness will not benefit the examination of the roughness values and surface characteristics of the plateau-honed surfaces [10].

Plateau honing provides significant reductions in peak roughness values and peak roughness material concentrations. With the plateau honing characteristic, the material density of a certain section becomes more stable and the surface acts more stable as a whole. This means that the risk of contact between the two parts working together is reduced, and contact occurs over a larger surface area, so that the surface pressure is decreased, and wear is reduced. However, the roughness values of the valley do not change, and its lubricant-supporting effect still continues [11].

Low oil consumption and a consequent reduction in harmful pollutant emissions from internal combustion engines are the principal objectives in the development of new techniques for honing piston raceways. These objectives can be met with slide honing, or, better still, helical slide honing. These three-stage processes are comparable with plateau honing and can be implemented on existing honing machines. Reducing the amount of oil in the combustion chamber is therefore important for reducing harmful emissions. Cross-hatch-honed cylinder bore surfaces have a particular part to play in this; they must hold a sufficient quantity of oil to prevent metal-to-metal contact between the friction pairs. On the other hand, excess oil has the detrimental effect on emissions.

Siqi Ma et al. demonstrated that the friction coefficient of the piston ring cylinder liner first decreased and then increased with the increasing

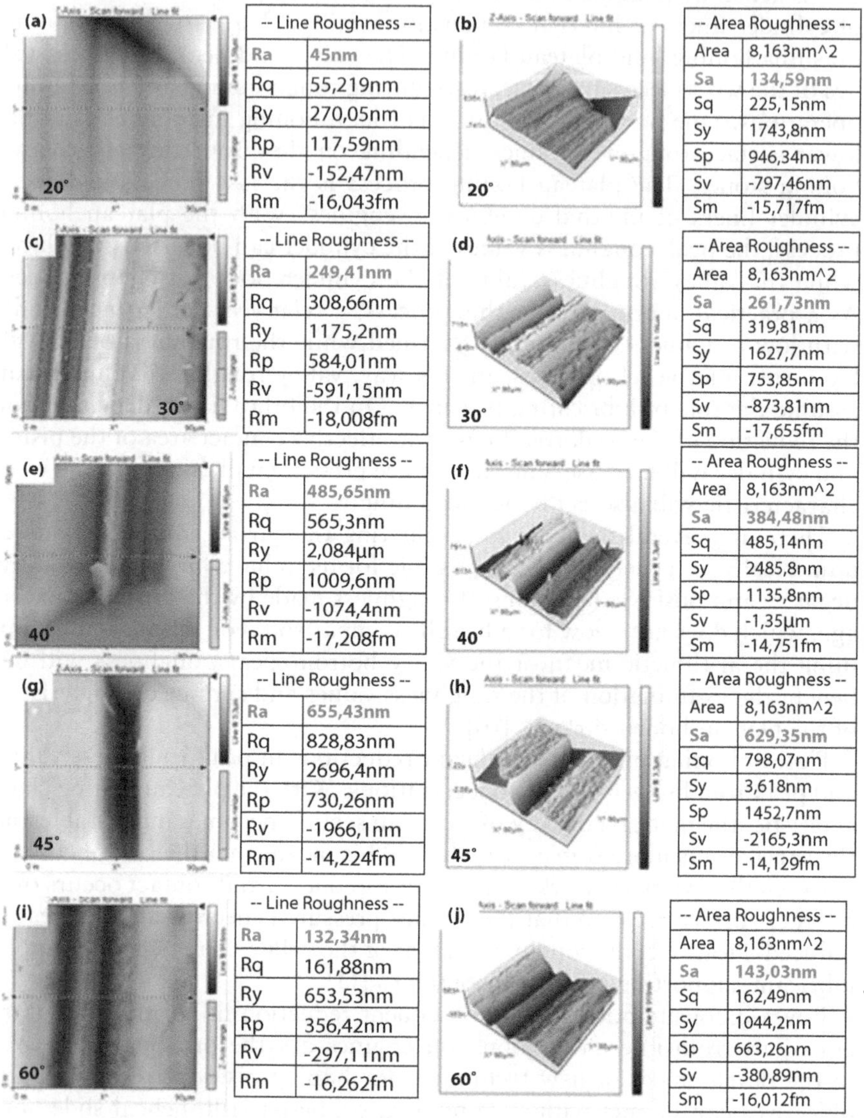

Figure 1.14 AFM roughness comparison of the balls tested on the 20°, 30°, 40°, 45°, 60° honed cylinder liners.

honing cross-hatch angles (from 47° to 65°), supporting our work. With the decrease in honing cross-hatch angle (from 65° to 58°), the direction of honing groove distribution tended to be perpendicular to the sliding direction. The honing orientation could form a better blocking effect on the lubrication oil flow during the sliding process of the friction pair, which contributed to the load-bearing capacity of the oil wedge [12].

Table 1.11 2D and 3D roughness values of the wear scar of the balls tested on 20°, 30°, 40°, 45°, and 60° honed cylinder liners

Honing angle	*Sa (nm)*	*Ra (nm)*
20°	**134.59**	**45**
30°	261.73	249.41
40°	384.48	485.65
45°	629.35	655.43
60°	143.03	132.34

1.3.2 Comparison of the AFM analysis of balls and honed cylinder liner after tribotests

Figure 1.14 presents 2D (a, c, e, g, i) and 3D (b, d, f, h, j) roughness images measured in atomic force microscopy (AFM) of the wear scar of the balls rubbed on the five honing angles (20°, 30°, 40°, 45°, and 60°) of the liners.

Table 1.11 shows 2D and 3D roughness values in nanometers of the 20°, 30°, 40°, 45°, and 60° honed cylinder liners. Roughness parameters such as Sa and Ra values of the wear scar diameter presented their lowest values in 20° honed liners according to the tested 30°, 40°, 45°, and 60° honed cylinder liners.

Figure 1.15 presents 2D (a, c, e, g, i) and 3D (b, d, f, h, j) roughness images measured by AFM of the wear track of the five honing angles (20°, 30°, 40°, 45°, and 60°) on liner specimens rubbed under a steel ball.

Table 1.12 shows 2D and 3D roughness value in nanometers of the wear scar of the 20°, 30°, 40°, 45°, and 60° honed cylinder liners. Roughness parameters such as Sa and Ra values of the wear scar presented the lowest values in 20° honed liner according to the tested 30°, 40°, 45°, 60° honed cylinder liners.

Chapter 1 has great importance for sustainability. The demands for decreased environmental impact from vehicles are resulting in a strong push for decreased engine oil and fuel consumption. Engine oil and fuel consumption are to a great extent controlled by the topography of the cylinder liner surface. Smooth honing liners represent the latest developments in commercial liner manufacturing and are the current state of the art. Optimum honing angles and well-polished roughness parameters are reducing fuel and oil consumption, increasing wear resistance.

1.4 CONCLUSIONS

1. The effect of honing angles such as 20°, 30°, 40°, 45°, and 60° applied on the cylinder barrel of diesel engines on friction and wear is well investigated using a reciprocating tribometer.

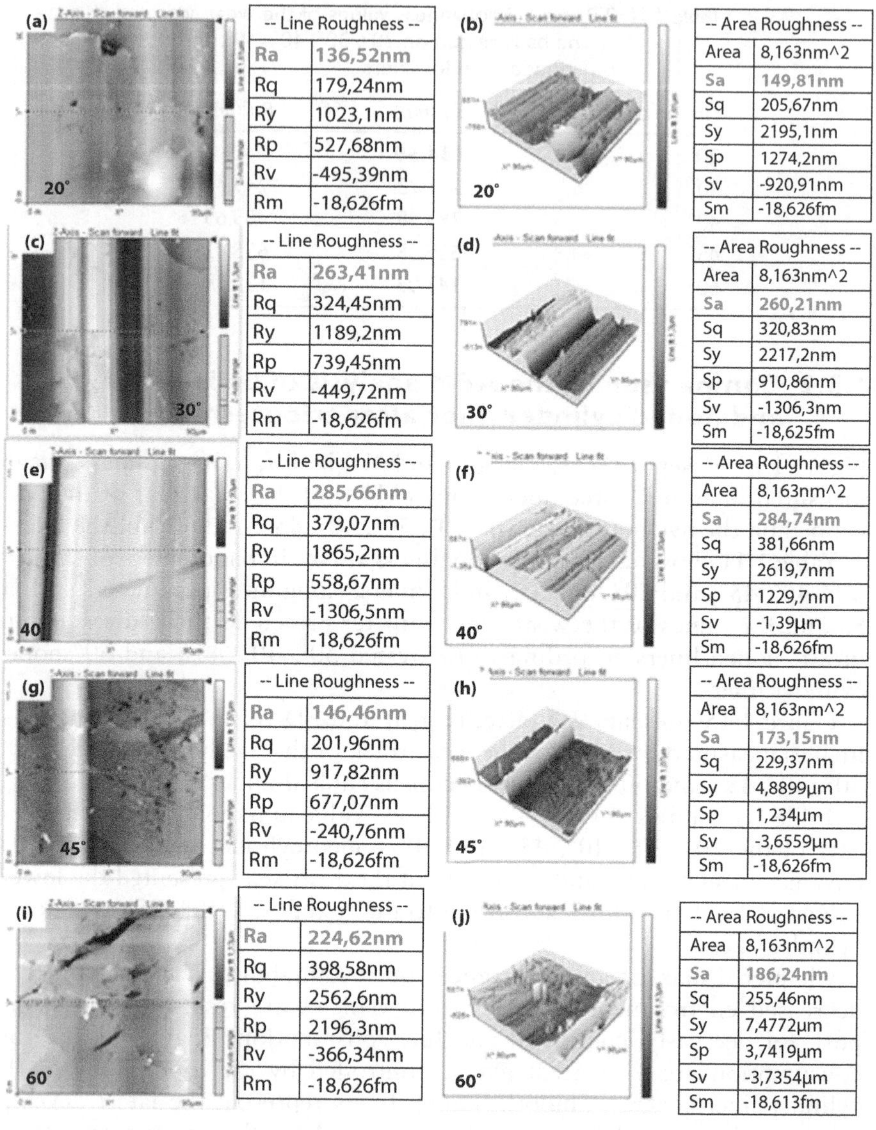

(a) 20°

-- Line Roughness --	
Ra	136,52nm
Rq	179,24nm
Ry	1023,1nm
Rp	527,68nm
Rv	-495,39nm
Rm	-18,626fm

(b) 20°

-- Area Roughness --	
Area	8,163nm^2
Sa	149,81nm
Sq	205,67nm
Sy	2195,1nm
Sp	1274,2nm
Sv	-920,91nm
Sm	-18,626fm

(c) 30°

-- Line Roughness --	
Ra	263,41nm
Rq	324,45nm
Ry	1189,2nm
Rp	739,45nm
Rv	-449,72nm
Rm	-18,626fm

(d) 30°

-- Area Roughness --	
Area	8,163nm^2
Sa	260,21nm
Sq	320,83nm
Sy	2217,2nm
Sp	910,86nm
Sv	-1306,3nm
Sm	-18,625fm

(e) 40°

-- Line Roughness --	
Ra	285,66nm
Rq	379,07nm
Ry	1865,2nm
Rp	558,67nm
Rv	-1306,5nm
Rm	-18,626fm

(f) 40°

-- Area Roughness --	
Area	8,163nm^2
Sa	284,74nm
Sq	381,66nm
Sy	2619,7nm
Sp	1229,7nm
Sv	-1,39µm
Sm	-18,626fm

(g) 45°

-- Line Roughness --	
Ra	146,46nm
Rq	201,96nm
Ry	917,82nm
Rp	677,07nm
Rv	-240,76nm
Rm	-18,626fm

(h) 45°

-- Area Roughness --	
Area	8,163nm^2
Sa	173,15nm
Sq	229,37nm
Sy	4,8899µm
Sp	1,234µm
Sv	-3,6559µm
Sm	-18,626fm

(i) 60°

-- Line Roughness --	
Ra	224,62nm
Rq	398,58nm
Ry	2562,6nm
Rp	2196,3nm
Rv	-366,34nm
Rm	-18,626fm

(j) 60°

-- Area Roughness --	
Area	8,163nm^2
Sa	186,24nm
Sq	255,46nm
Sy	7,4772µm
Sp	3,7419µm
Sv	-3,7354µm
Sm	-18,613fm

Figure 1.15 AFM roughness comparison of the wear track of 20°, 30°, 40°, 45°, and 60° honed cylinder liners.

2. It has been demonstrated that the wear scar diameter presented the lowest value horizontally and vertically in the 20° honed angle compared to 30°, 40°, 45°, and 60° honed cylinder liners in digital optical microscopic examination.

Table 1.12 2D and 3D roughness value of the wear track of the 20°, 30°, 40°, 45°, and 60° honed cylinder liners

Honing angle	*Sa (nm)*	*Ra (nm)*
20°	**149.81**	**136.52**
30°	260.21	263.42
40°	284.74	285.66
45°	173.15	146.46
60°	186.24	224.62

3. It has been demonstrated that the wear scar distance of the 20° honed angle presented the lowest value compared to 30°, 40°, 45°, and 60° honed cylinder liners in digital optical microscopic examination.
4. It has also been found that the roughness parameters such as Sa and Ra values decreased from 60°, 45°, 40°, and 30° to 20° on the *non-tested* honed liner in optical digital microscopic examination.
5. It has been demonstrated that the roughness parameters such as Sa and Ra values of the wear scar diameter presented their lowest values in the 20° honed liner according to the tested 30°, 40°, 45°, and 60° honed cylinder liners in optical digital and atomic microscopic examination.
6. It has been demonstrated that the roughness parameters such as Sa and Ra values of the wear scar distance presented their lowest values in 20° honed liner according to the tested 30°, 40°, 45°, and 60° honed cylinder liners in optical digital and atomic force microscopic examination.
7. At the end of the tribotests, roughness parameters such as Sa and Ra values decreased because there is a homogeneous plastic deformation that makes the surface smooth and reactive, providing better surface degradation. This allows quick surface reaction between the lubricants and sliding pairs of the 20° cross-hatch honed liner. Then additives such as Zn, Ca, and P content formed in the protective layers and reduced the coefficient of friction.

ACKNOWLEDGMENTS

The authors thank Mr. Zeki Ayaz, director of Anadolu Motor of the Anadolu Group, for providing spare parts and the honing process for the PhD work of Veli Uysal.

REFERENCES

1. Dimkovski, Z., Anderberg, C., Ohlsson, R., Rosén, B.G. 2011. "Characterisation of worn cylinder liner surfaces by segmentation of honing and wear scratches", *Wear*, 271, 548–552.
2. Vrac, D., Sidjanin, L., Balos, S. 2014. "The effect of honing speed and grain size on surface roughness and material removal rate during honing", *Acta Polytechnica Hungarica*, 11, 10, 163–175.
3. Klocke, F. 2009. *Manufacturing Prosess 2 Grinding, Honing, Lapping (RWTH edition)*. Deblik, Berlin: Springer.
4. Kısa, M. 2002. *Özel Üretim Teknikleri*. Bursa: Furkan Ofset.
5. Nirmal Raj, L., Rajasaker, K., Vetriselvam, M., Vignesh, R., Suresh, S. 2017. "Straight pass rate improvement of productivity in honing process", *International Journal of Innovative Research in Science, Engineering and Technology*, 6, 3, 4718–4724.
6. Uhlmann, E., Spur, G., Kleinschnitker, M. 2015, Bölüm 5, Sayfa 234. Honing and superfinishing. *Handbook of Ceramics Grinding and Polishing*. Editör: Doi, T., Marinescu, L.D., Uhlmann, E. Elsevier Tokyo, Japan.
7. Bouassida, H., Biboulet, N., Sainsot, P., Lubrecht, A.A. 2014. "Piston ring load carrying capacity: Influence of cross-hatching parameters", *Proceedings of the Institution of Mechanical Engineers, Part J: Journal of Engineering Tribology*, 228, 6, 642–648.
8. Biboulet, N., Bouassida, H., Lubrecht, A.A. 2015. "Cross hatched texture influence on the load carrying capacity of oil control rings", *Tribology International*, 82, 12–19.
9. Demirtas, S., Kaleli, H., Khadem, M., Kim, D.E. 2019. "Characterization of the friction and wear effects of graphene nanoparticles in oil on the ring/cylinder liner of internal combustion engine", *Industrial Lubrication and Tribology*, 71, 642. DOI: 10.1108/ILT-05-2018-0170.
10. Lipa, Z., Tomaníckovǻ, D. 2011. "Utilisation of Abbott-Firestone curves characteristics for the determination of turned surface properties", *Annals of the Faculty of Engineering Hunedoara*, 9(3), 223.
11. Corral, I.B., Calvet, J.V., Salcedo, M.C. 2010. "Use of roughness probability parameters to quantify the material removed in plateau-honing", *International Journal of Machine Tools and Manufacture*, 50, 7, 621–629.
12. Siqi, M., Yuchen, L., Zichun, W., Zheng, W., Ruoxuan, H., Jiujun, X. 2019. "The effect of honing angle and roughness height on the tribological performance of CuNiCr iron liner", *Metals*, 9, 487.

Chapter 2

Effect of surface textures of friction plates on wear behaviour and torque transmission in compressors with clutch mechanism

Mehmet Esat Aydin, Esathan Bayraktar, Erdem Unuvar, and Mehmet Bagci

2.1 INTRODUCTION

Compressed air is needed in vital components such as brakes and suspension in heavy vehicles and/or in the drive mechanisms such as bellows and door opening/closing. The necessary compressed air is produced by compressors that are connected to the internal combustion engine in the vehicle. Compressors send compressed air that is produced to compressed air storage tanks on the vehicle, and from the tanks it is transmitted to the places of use (brakes, suspension, door-opening mechanism, bellows, etc.). After the tanks are filled, since the compressor is constantly connected to the vehicle's engine, it produces unnecessary air, resulting in constant power consumption. Clutch mechanisms have been developed to prevent unnecessary power consumption and save energy.

There are many types of clutches in the literature, such as rigid, flexible, frictional, dissolvable, electrostatic, or lamellar. In cases where the compressor starts to work unnecessarily, the clutch mechanism is activated, disconnecting the compressor and motor or disengaging, allowing the compressor and motor to reconnect. Consequently, unnecessary power consumption is prevented. Multi-lamellar clutches, which are also the subject of this study, keep the start-up and stop-out times as short as possible, and thus the compressor can be quickly activated or deactivated [1].

The motion and torque received from the engine are transferred to the lamellae, also called friction plates, through a gear. The opposing lamellae, which come into contact with the friction plates and surface to surface, start to rotate under the influence of friction and transmit the resulting movement to the compressor. In other words, it can be said that there is a pair of lamellae connected to each other surface to surface, and the resulting movement is transmitted through friction [2]. The surface of the friction plates or lamellae can be wet or dry. Accordingly, the wear behaviour of the clutch may vary.

DOI: 10.1201/9781003713364-2

When the clutch mechanism is activated, the lamellae pairs are separated from each other, and then when the clutch is disengaged, contact occurs again. Over time, wear is observed between the lamellae that try to move relative to each other by friction [3]. Due to wear on the surfaces of the friction plates, the desired locking doesn't occur between the pairs of lamellae. Therefore, movement and torque transfer don't continue properly. In this movement transfer, lamellar surfaces are required to be rough. However, when too much roughness occurs, the plates are exposed to excessive adhesive wear depending on the load acting on them. In addition, heat transfer between heated surfaces due to friction also creates an effect on deteriorating the surfaces. There are many academic studies on lamellar clutches or multi-disc clutches in the literature. Many issues such as the wear pattern developed by the lamellae depending on operating conditions, surface conditions, torque transmission, and the effect of material or lubricant on friction have been examined.

In one study, composite friction materials that can be used in brake and clutch applications were investigated. When the clutch and brake mechanism are activated, it was examined how the material will behave against wear, taking into account the production method [4]. The friction situation that occurs during the contact of a paper-based friction material and glass fibre material used in the clutch mechanism under different pressures was examined by Ingram et al. [5]. They studied the wear caused by load and obtained images with various surface measurement methods. They concluded that as the load increases, plastic deformations increase, and therefore the amount of wear increases. In a different study, torque levels and coefficient of friction values were measured by adding various oils to a friction plate clutch mechanism, and the effect of oils on these parameters was examined [6]. During the short-term contact moments in a multi-disc clutch, it has been observed that plates rubbing against each other create hot spots. In their study, Zagrodski and Truncone took into account the clutch of a heavy vehicle and included numerical investigations. The hot spot formation of two different friction materials and different thicknesses of plates was investigated, and they found that micro defects in the discs have a significant effect on hot spot formation [7]. In another study, it was found that the heat generated by friction in a multi-disc clutch creates significant thermal stress [8]. Therefore, the resulting thermal stresses exhibit behaviour to degrade the surfaces. Similar to the subject investigated in this study, Nyman et al. investigated the effects of friction on the surface and surface topography of the discs of a clutch mechanism. Sintered material was used in the study, and the condition of the surfaces was analysed depending on the operation of the clutch. Consequently, the surface topography of the material was found to affect the frictional properties of the grip, and a method was developed to calculate the wear properties from the topography [9].

In the literature, there are many studies investigating friction plates in multi-disc clutches from various aspects. As one of the different subjects, a method for detecting the failure of the multi-disc clutch mechanism has been developed [10]. It was determined that the torque acting on the discs in the mechanism, friction, and wear on the plate could be the sources of failure. When looking at the results, it was observed that abrasive wear, ploughing wear, adhesive wear, and delamination wear mechanisms occurred on the discs between 15°C and 400°C, and this result is proportional to the results found by Scott and Suntiwattana [6, 10]. In another study, a wet friction plate clutch was modelled and analysed. A model was developed to determine the dynamic behaviour and vibration characteristics of the clutch. In addition, viscous effects due to lubrication were also taken into account in the study. It was interpreted that the vibration caused by the loads acting on the clutch mechanism does not affect the natural vibration [11].

A study was conducted on the clutch mechanism with paper-derived plates, which is also the subject of this study by Natsumeda and Miyoshi, and in the study, a theoretical analysis of the gripping process of paper-based wet plates was made. In addition, a hydrodynamic lubrication model was created and the effect of lubrication was investigated. Moreover, the effect of heat generated by friction on the viscosity decrease has been modelled mathematically [12]. In addition, equations were created for torque, a comparison was made with the measured torque values, and it was concluded that the results were proportional to each other. In another study, an attempt was made to determine the friction and wear characteristics of a wet clutch with copper (Cu)-containing friction plates. The effect of surface roughness level on wear was investigated. Consequently, it has been determined that roughness and thermal effects have a friction-increasing effect [13].

Mansouri et al. determined the dynamic and thermal characteristics of the wet clutch by applying torque to a clutch mechanism. This study has been elaborated in some detail; the clutch was exposed to different types of lubrication, and the coefficient of friction was measured. Additionally, the data was taken by applying torque to the structure at a level that occurs in the vehicle [14]. In another study, the torque transmission in the clutch mechanism and the effect of friction on torque transfer were investigated. The measured torque values were higher than the theoretical torque values, and therefore, it is understood that friction slows torque transmission. It has also been advocated that the rubbing surfaces should be accustomed to each other [15]. Friction plates in a wet clutch mechanism were examined with a pin-on-disc tribometer set, and the temperatures occurring during friction were measured by Marklund and Larsson [16]. In this way, the friction behaviour of the clutch and the effect of friction on torque were determined. When looked at in the literature, issues such as friction and wear behaviour of lamella or disc clutches, the effect of friction on torque transfer, and the effect of lubrication on friction performance come to the fore.

Another factor that affects friction is the material used. In a published study, it was attempted to determine the friction stability of carbon fibre reinforcement friction material depending on the operating conditions. The material was used in a brake mechanism, and according to the results, it was determined that the effect of the material on braking was affected by pressure, speed, oil temperature, and flow rate [17]. In another study, the effect of carbon fibre/silicon carbide (C/SiC) friction material on braking was examined, and the C/SiC material was compared to a conventional carbon fibre or ceramic-derived material by applying various wear values. According to the results, while the friction coefficient of the C/SiC material was higher than that of classical carbon fibre, it was equal to that of the ceramic material [18]. In another study, the friction properties and oil retention of sintered bronze materials and paper-based materials used in wet clutches were measured [19]. It has been understood that sintered parts can be used in the clutch mechanism. In another study, the tribological properties of sintered friction material (Fe_2O_3) and steel discs used in the brake mechanism were tried to be determined comparatively [20]. The performance of materials was measured according to braking at 8 m/s speed and variable pressures. The morphology of the surfaces, SEM images, x-ray images, and friction coefficient were measured. According to the results, while the friction coefficient of the two materials is very different, the friction coefficient value of both materials increased proportionally depending on the thickness and load.

In this study, the components of the multi-disc clutch mechanism produced by the original equipment manufacturer (OEM) and the clutch mechanism of YUMAK, which produces aftermarket products, will be compared. The difference between OEM and aftermarket is herein described: When any company manufactures a high-duty vehicle, the company that it has contracted to produce a component (e.g. compressor) in the vehicle is the OEM. The companies that produce the products of OEM companies with reverse engineering studies are known as aftermarket. YUMAK represents the aftermarket company in this context. In this study, friction plates in the clutch mechanism of two different companies will be compared in terms of wear properties, friction torque, surface topography before and after wear, and microscope images.

2.2 METHODOLOGY

2.2.1 Preparation of sample

Movement and torque transmission occur when two different friction plate surfaces rub against each other. Both plates have a carbon friction surface but differ from each other in terms of pattern. It was also observed that carbon paper surfaces are suitable for friction in the study conducted by Fei

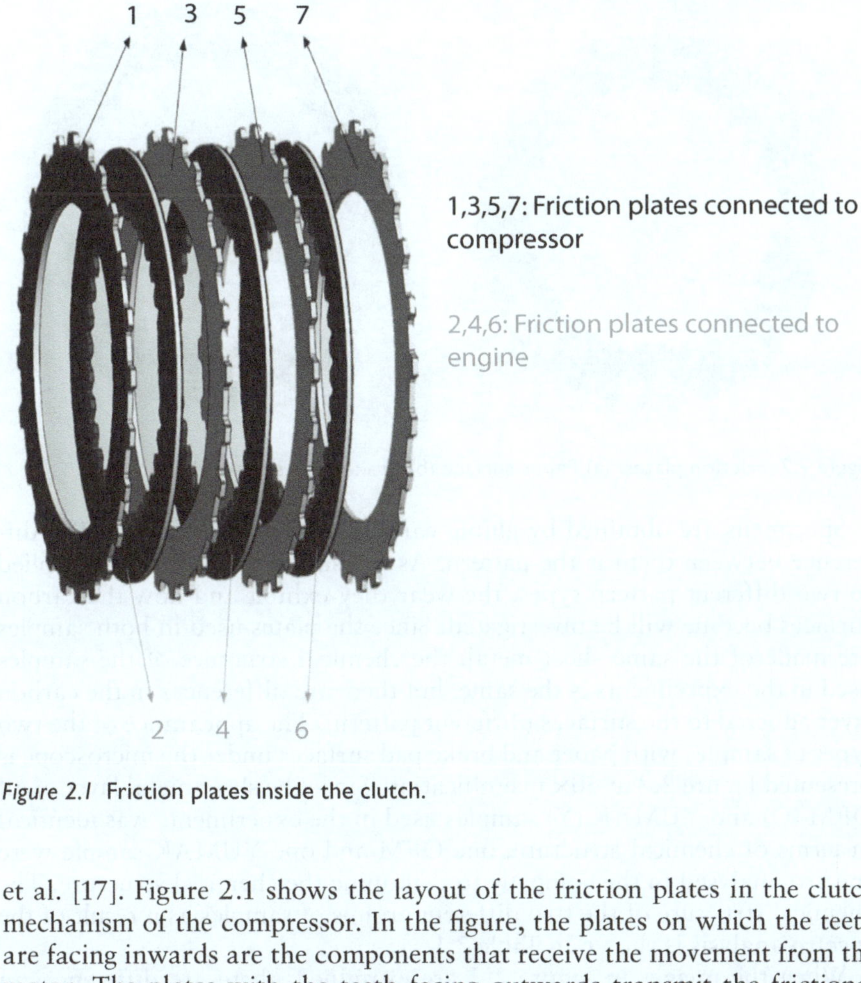

Figure 2.1 Friction plates inside the clutch.

et al. [17]. Figure 2.1 shows the layout of the friction plates in the clutch mechanism of the compressor. In the figure, the plates on which the teeth are facing inwards are the components that receive the movement from the motor. The plates with the teeth facing outwards transmit the frictional movement of the crank of the compressor. As a result of the surface-to-surface contact of two different plates, motion transfer occurs due to the friction effect. The disadvantages of being frictionally connected to each other are that the surfaces start to wear out after a certain period of time. Two types of friction plates with paper and brake pad surfaces are shown in Figure 2.2.

Wear tests will be applied to both specimens under various loads and speeds in three repetitions. As a result of the tests, it will be determined what kind of wear and surface behaviour the paper and brake pad structures will exhibit compared to the various differences it has.

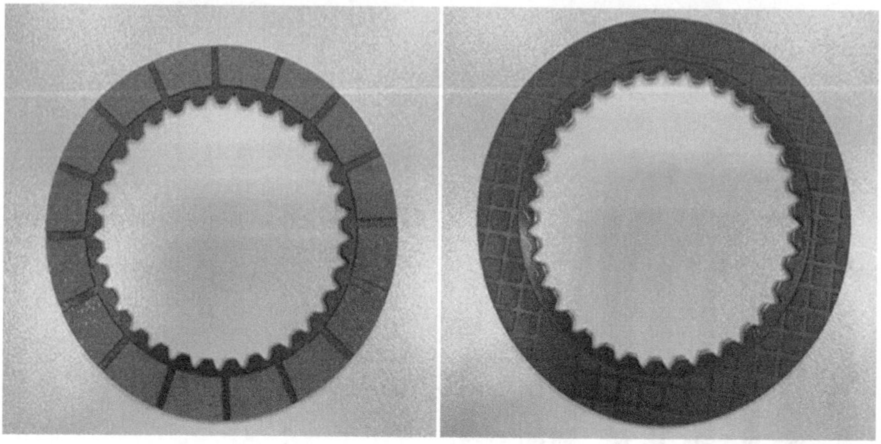

Figure 2.2 Friction plates, (a) Paper surface (b) Brake pads surface.

Specimens are obtained by gluing carbon paper to the surface. The difference between them is the pattern. As a result of the wear tests applied to two different pattern types, the wear they exhibit and how the carbon surfaces become will be investigated. Since the plates used in both samples are made of the same sheet metal, the chemical structure of the samples used in the experiments is the same, but there are differences in the carbon layer adhered to the surfaces (different pattern). The appearance of the two types of samples with paper and brake pad surfaces under the microscope is presented Figure 2.3 at 50× magnification. Since the sheet metal layer of all OEM (O) and YUMAK (Y) samples used in the experiments was identical in terms of chemical structure, one OEM and one YUMAK sample were spectro analysed in the company to determine the chemical structure. The chemical structure of the two different groups of samples as a result of the spectro analysis is shown in Table 2.1.

When the images in Figure 2.3 are examined, there are differences in the paper and brake pad surface specimens of two different companies. In OEM specimens, the resin density is more dominant and therefore results in a more blurred image under the microscope. Among the OEM paper and brake pad surface specimens, the fibres in the carbon layer with the brake pad surface were cropped and not obvious; long fibres were used in the paper surface specimen. Among the YUMAK samples, long fibres were used in both types of specimens, but the fibre density is less on the paper surface. The fibre type is also different between the OEM brake pad surface and YUMAK brake pad surface. Among paper surfaces, the YUMAK sample has both a lower resin density and lower fibre density compared to OEM. It will be discussed by looking at the surface conditions after wear.

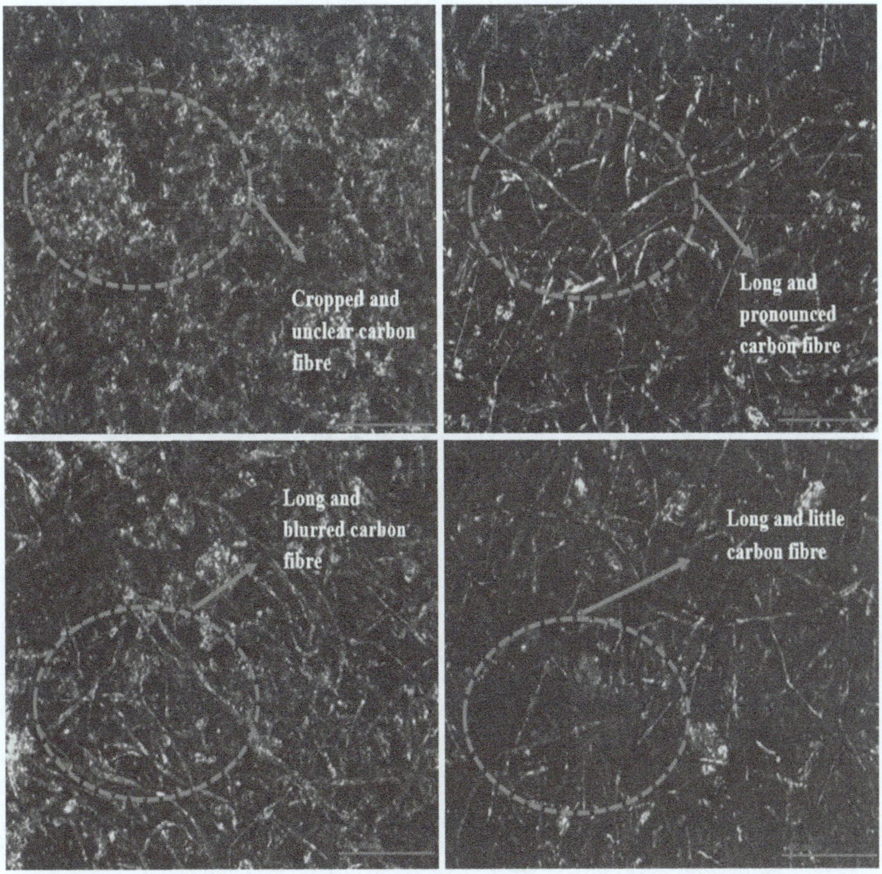

Figure 2.3 Microscope images, (a) OEM brake pad surface – O1, (b) YUMAK brake pad surface – Y1, (c) OEM paper surface - O2, (d) YUMAK paper surface - Y2.

It can be seen from Table 2.1 that both structures are dominated by iron and are almost identical. The experimental device used to determine the coefficient of friction on the specimens was made with a pin-on-disc set in accordance with ASTM G99. The visualisation of the test unit is given in Figure 2.4.

In the device shown in Figure 2.4, variable loads and speeds were taken into consideration while measuring the coefficient of friction. In terms of contact times, measurement was made for 120 seconds. The changes in loads and speeds occur when the clutch mechanism is operating on the compressor and the load values acting on the friction plates change as the clutch is activated and deactivated. Consequently, in order to simulate the variation of friction torque with respect to thrust force and speed parameters, two different experiments were set up by varying the load and speed values.

Table 2.1 Chemical structures of OEM (O) and YUMAK (Y) samples

Sample	*Element information found in the structure*										
	Fe (%)	*C (%)*	*Si (%)*	*Mn (%)*	*P (%)*	*S (%)*	*Cr (%)*	*Mo (%)*	*Ni (%)*	*Cu (%)*	*Sn (%)*
YUMAK	98.9	0.157	0.252	0.534	0.0149	0.0061	0.0542	0.0043	0.0216	0.0137	0.0029
OEM	98.9	0.157	0.249	0.542	0.0145	0.0043	0.0538	0.0042	0.0193	0.0131	0.003

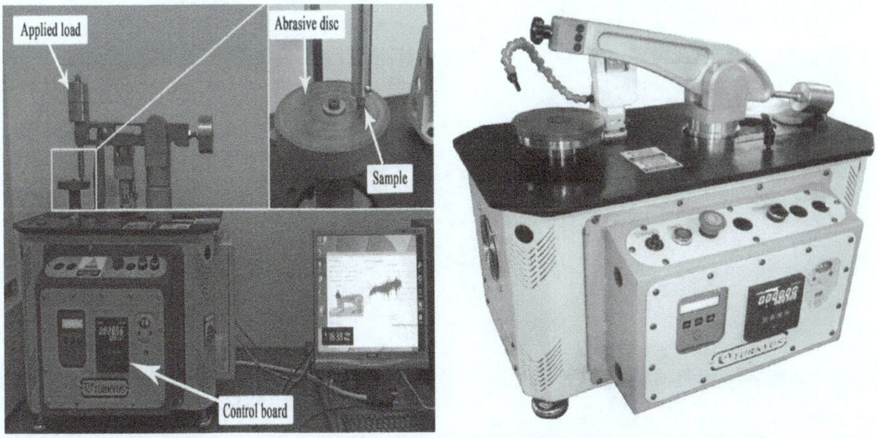

Figure 2.4 Pin on disc tribometer device, (a) Components information, (b) Device appearance.

Table 2.2 Experiment parameters

Experiment	*Load (N)*	*Speed (RPM)*	*Time (seconds)*
1	20	250	120
2	45	400	120

Since the clutch engagement and disengagement time is approximately 120 seconds (it is known by the company), this value was preferred for the test time. The parameters used in the experiments are shown in Table 2.2.

The samples used in the pin-on-disc set are shown in Figure 2.5. Surface patterns of Y1 and O1 appear on both specimens. However, paper-patterned Y2 and O2 specimens have a more uniform appearance.

The coefficient of friction values for four specimens were measured graphically according to the experiments given in Table 2.2. In addition to the friction coefficient, microscope images of the samples before and after wear were also observed at 50× magnification. Moreover, in order to observe the condition of the OEM (O) and YUMAK (Y) friction plate surfaces in detail after wear, surface topography was measured and 3D mapping was performed. The experiments were carried out with three repetitions; the increase in load and velocity was investigated in terms of various surface behaviours of the structure.

Another important issue in friction discs is the friction torque. In this context, friction torque values were measured in relation to speed in order to understand how much the friction plates create an opposing effect against the rotational motion transmitted in the clutch mechanism depending on load and speed.

Figure 2.5 Test Specimens before wear test, Before wear test specimens (a) Y1, (b) O1, (c) Y2, (d) O2.

2.3 RESULTS AND DISCUSSION

2.3.1 Coefficient of friction results

Wear tests were performed on the pin-on-disc set of the friction plates with two separate test parameters. The pin-on-disc wear test was applied because the friction plates are connected to each other surface to surface within the clutch mechanism and a linear contact occurs between them. As a result of the tests, the friction coefficient values depending on time were measured. The adhesion ability of the plates to each other should be interpreted depending on the friction that occurs during surface-to-surface contact. Therefore, as a result of the tests, the resulting friction coefficient change was determined. The friction results applied to four different specimens are shown in Figure 2.6 and Figure 2.7. The friction changes in the

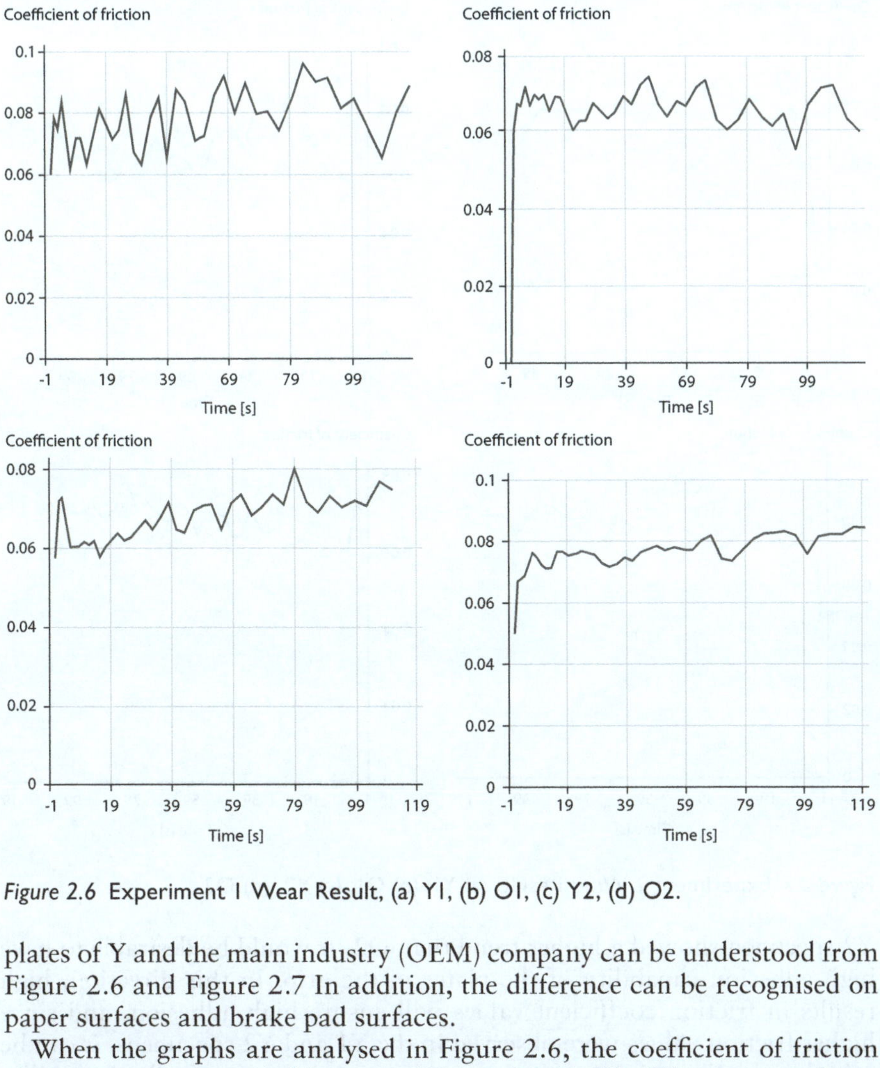

Figure 2.6 Experiment I Wear Result, (a) Y1, (b) O1, (c) Y2, (d) O2.

plates of Y and the main industry (OEM) company can be understood from Figure 2.6 and Figure 2.7 In addition, the difference can be recognised on paper surfaces and brake pad surfaces.

When the graphs are analysed in Figure 2.6, the coefficient of friction change intensively varied between 0.06 and 0.08. The Y1 specimen shows results significantly exceeding 0.08. The values generally started at 0.06 and increased towards 0.08. In addition, the lapping-in moment is more evident in OEM samples. According to the first experiment performed under low load and speed, the friction coefficient values of the specimens did not show much variation. While paper surfaces give similar results to each other, it can be said that several differences occur between the brake pad surfaces. The Y1 surface exhibited a gradually increasing frictional behaviour and a higher amount of variability, while O2 exhibited a more stable structure. However, no decrease in friction values was observed. When evaluated in terms of adhesion results, the amount of variability in the Y1 sample was higher than in O1. Similarly, the change in the coefficient of friction of the

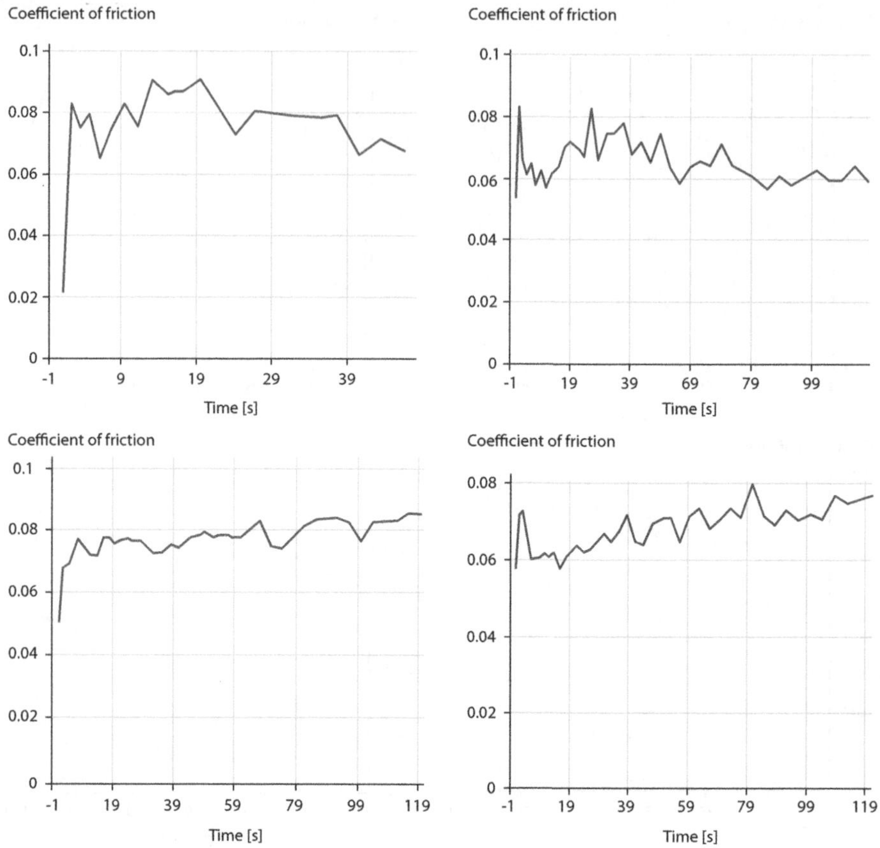

Figure 2.7 Experiment 2 Wear Result, (a) Y1, (b) O1, (c) Y2, (d) O2.

Y2 specimen showed a higher trend than O2. It would be desirable to have high adhesion capability of the plates in the grip. In this direction, high results in friction coefficient values will ensure high adhesion, and since higher friction values were observed in the Y1 and Y2 specimens, it can be said that the Y specimens are more appropriate in terms of adhesion ability compared to OEM.

In the second experiment, in which load and speed values were increased, similar graphs are shows in Figure 2.7. When Figure 2.7 is analysed, firstly, the Y1 surface was destroyed during the test and the experiment was terminated at 45 seconds. Based on the 45 seconds of friction results, a very distinct and very unstable structure has been formed at that moment. The O1 surface has a more stable structure and exhibited a decreasing behaviour in terms of friction. When the graphs obtained in the second experiment performed at high load and speed are evaluated in terms of adhesion ability, unlike the first experiment, the friction coefficient change in the O1 specimen was higher. Therefore, it can be said that the adhesion ability of the O1 type sample increases at high loads and speeds. In addition, Y1 type specimens disintegrate under high load and speed even if they have high adhesion. Therefore, they are not very suitable. Paper surface type O and Y specimens showed similar behaviour despite the increase in load and speed.

When load and speed values are low (Figure 2.6), the coefficient of friction was measured between 0.06 and 0.08 on average, but when load and speed values were increased (Figure 2.7), friction changed from 0.08 to 0.06. If the comparison between paper surfaces is examined, both specimens (Figure 2.6 [c,d]) exhibited increasing friction behaviour. There isn't a very high variation in terms of variability. In addition to these, between the two experiments, the increase in load and speed values did not change the coefficient of friction much but some changes have occurred in terms of friction behaviour. Moreover, the Y1 surface sample could not withstand the load and speed. Specimen images are shown after the wear test in Figure 2.8.

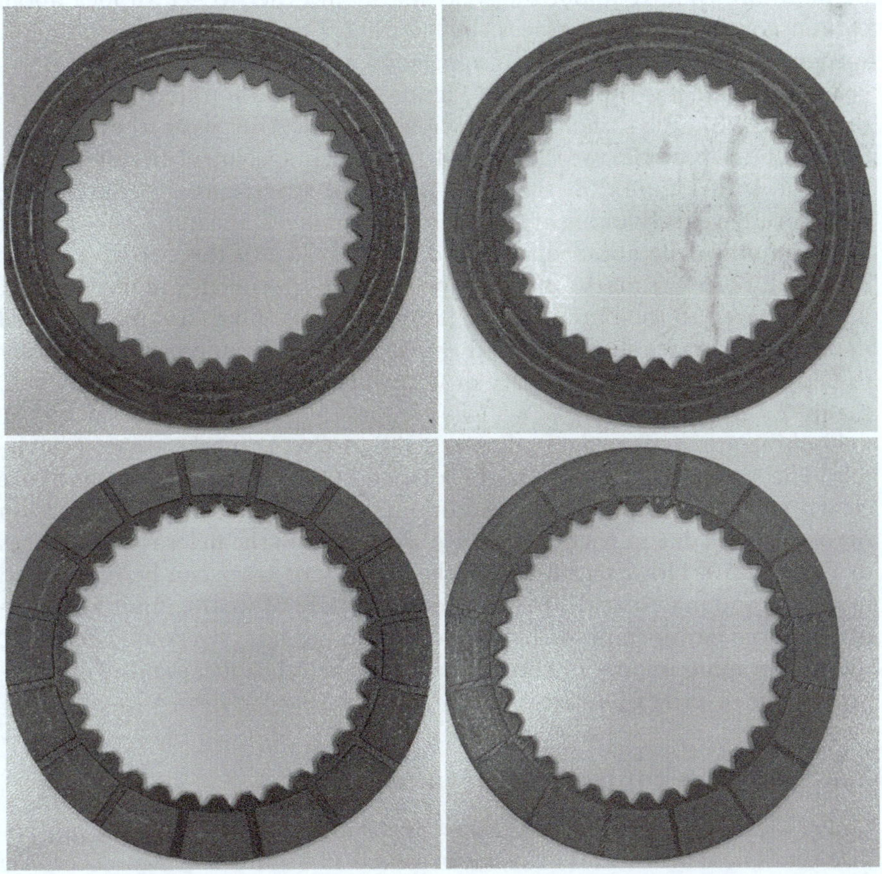

Figure 2.8 Test specimens after wear tests, (a) Y1, (b) O1, (c) Y2, (d) O2.

When the post-wear images are examined, the patterns disappear in the worn areas of the specimens with brake pad surfaces. In addition, the Y1 surface specimen can be recognised from Figure 2.8a as being more irritated. The wear marks are not as pronounced on paper-faced specimens (2.8 [c,d]) as on brake pad-faced surfaces. However, the most accurate interpretation can be made by surface topography and microscope images.

2.3.2 Microscope image results

Since carbon fibre-derived paper is adhered to the specimen surfaces, it is possible to see to lamellae originating from the fibres under the microscope image on the surfaces. When looking at the operating logic of the pin-on-disc set, wear is caused by the contact of a pin with a rotating disc. Therefore, a linear contact is formed, and these linear contact traces can be seen in Figure 2.8. However, microscope and macro-scale images will be more informative about the wear marks in order to get an idea about carbon coating on the surfaces. In Figure 2.9, microscope images of the specimens before and after wear are shown at 50× magnification.

When looking at the images in Figure 2.9, the carbon fibre particles on the surfaces of each sample are shown as dashed. After wear, these discontinuous fibre particles were removed from each specimen. It can be said, together with Figure 2.8, that Y1B and O1B specimens experienced the most severe wear. Looking at the microscope images, the fibres have almost disappeared in the abraded section on the surface of the two specimens. Although the wear marks on the paper surface specimens are not so obvious compared to the brake pad surface specimens, fibres are not observed in the paper surface specimens in the images under the microscope. In order to better understand the severity of wear, macro-scale images are given in Figure 2.10.

Looking at macro-images taken at 1300 µm, wear marks are clearly visible. The strong wear marks on the Y1B and O1B surface specimens b) are clearly visible on the surfaces. In addition, the deterioration and destruction of the surfaces due to friction can be observed from the macro-scale images in Figure 2.10. However, although the severity of wear can be recognised in micro- and macro-scale images, the magnitude of wear cannot be recognised. Some parameters, such as wear depth, are not observed. Therefore, the topographic images of the surfaces will be more meaningful in interpreting the severity of wear.

2.3.3 Surface topography results

Linear abrasion appeared on the specimens after friction with the pin-on-disc set. These worn areas can be observed in Figure 2.8. Moreover, the micro- and macro-scale images in Figure 2.9 and Figure 2.10 both illustrate and support this wear severity. However, none of these figures provide a

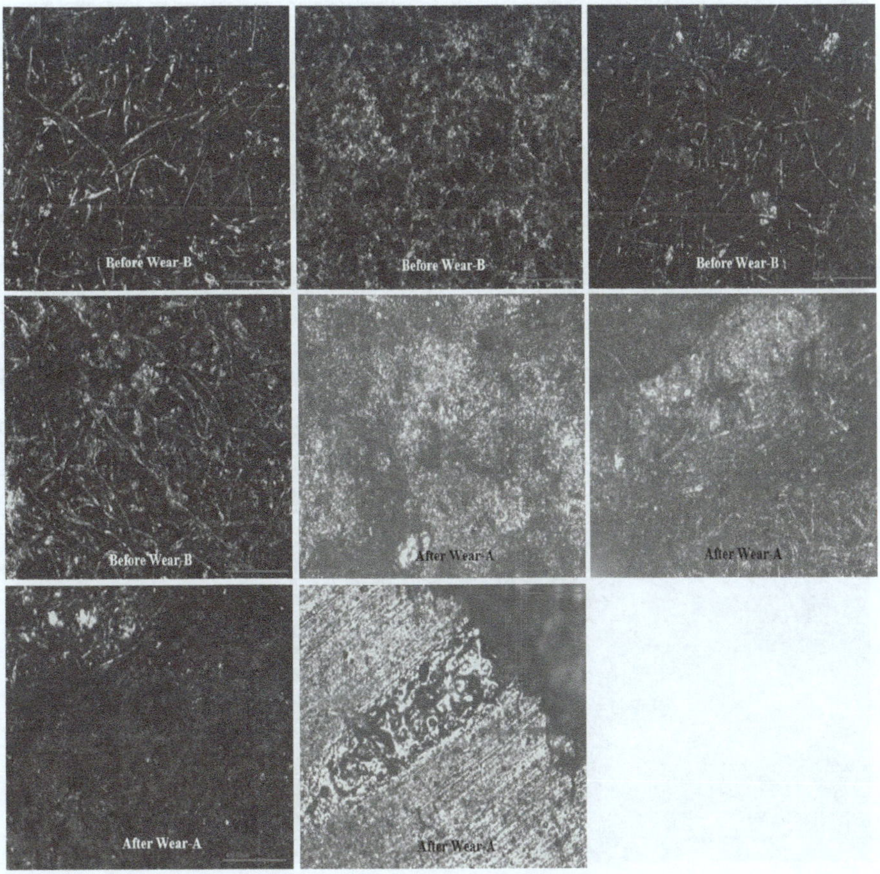

Figure 2.9 Microscope (50x) images, (a) Y1B-before wear(B), (b) O1B-before wear, (c) Y2B-before wear, (d) O2B-before wear, (e) Y1A-after wear(A), (f) O1A-after wear, (g) Y2A-after wear, (h) O2-after wear.

numerical representation of the magnitude of wear. An interpretation can only be made from the friction coefficient values in Figure 2.6 and Figure 2.7. In this direction, topographic images of the surfaces will both help us to understand the severity of wear and provide us with a numerical value.

Firstly, the topographical images of the Y and O specimens with brake pad surfaces are shown in Figure 2.11.

Topography measurements were taken over an area of 2 mm × 2 mm. Roughness values were also measured during the topography, but roughness values vary greatly from region to region because the wear areas on the surfaces are not homogeneous. Therefore, roughness values in Figure 2.11 are not very important. When looking at the surface images of the Y1A and O1A specimens, the wear pitting caused by two experiments on the Y specimen (Figure 2.11a) is also evident. However, the O specimen (Figure 2.11b)

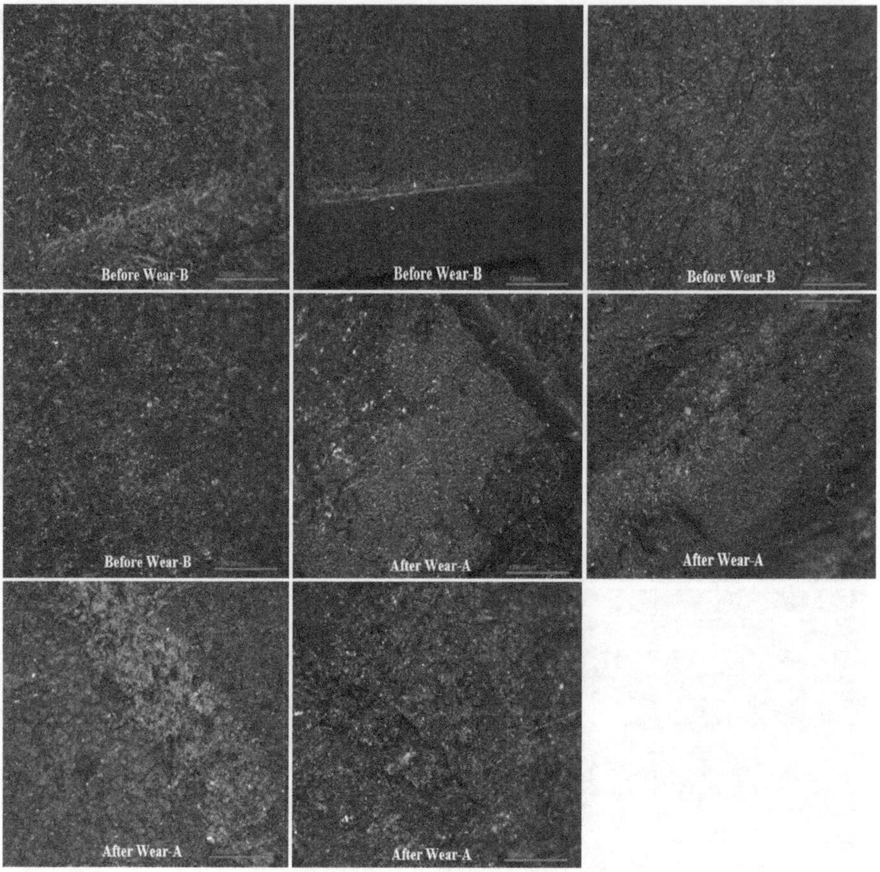

Figure 2.10 Macro-scale images, (a) Y1B, (b) O1B, (c) Y2B, (d) O2B, (e) Y1A, (f) O1A, (g) Y2A, (h) O2A.

has a single wear depth because it was imaged from a 2×2 area. In the second experiment, in which the load and speed values were increased, the Y specimen was subjected to very severe wear and the test was terminated at 45 seconds. The resulting wear images on the Y specimen can be seen in Figures 2.7–2.10 and 2.11a.

When evaluated numerically, if the non-wear surface of the O specimen Figure 2.11 (b) is considered as the 200 μm limit, the deepest zone after wear is reduced to 50 μm. In addition, the linearity of the wear pattern on the O specimen Figure 2.11 (b) can also be recognised from Figure 2.11b. Again, if 200 μm is taken as the non-wear surface reference, the area measured from the Y specimen Figure 2.11 (a) represents the inside of the wear section. Moreover, the depth of the wear zone decreased by up to 25 μm for the first test, while it decreased by up to 10 μm in the second test and the Y

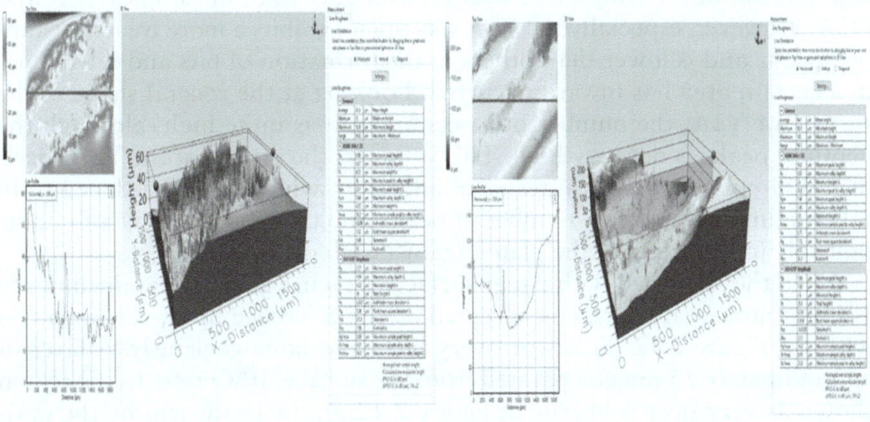

Figure 2.11 Surface topography, (a) Y1A, (b) O1A.

specimen Figure 2.11 (a). Previous examinations showed that the wear of the Y specimen was severe. Numerically, after wear, the specimen with the Y1A pitted approximately 30 μm more than the O Figure 2.11 (b). In terms of wear values, it shows that the specimen with O1A is more resistant to wear.

In the other direction, when looking at the wear images of the specimens (Figure 2.11), the number of peaks and pits in a given area was higher in the Y specimen (Figure 2.11a). Therefore, although the Y specimen is less resistant to abrasion, it can be said to be more superior in terms of adhesion ability.

The post-wear images of the Y2A and O2A specimens with paper surface are shown in Figure 2.12. Looking at the colour scales in the figures, the

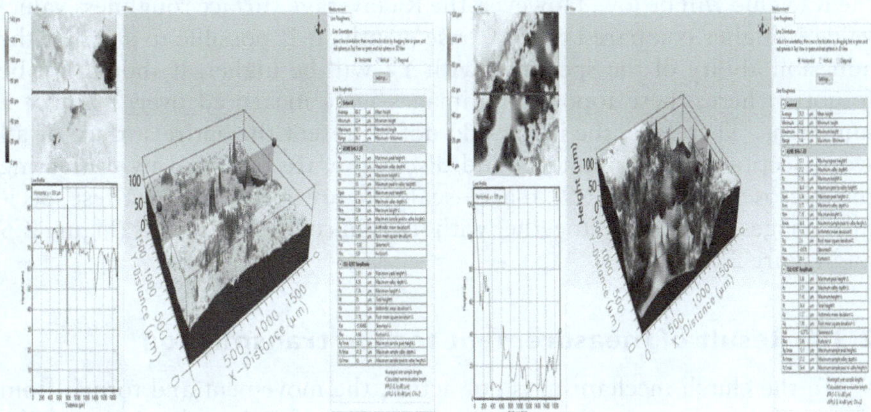

Figure 2.12 Surface Topography, (a) Y2A, (b) O2A.

depth variation in samples Y1 and O1 is higher than in samples Y2A and O2A. However, especially the Y2A specimen exhibits a more wear-resistant behaviour and is lower than others in the formation of pits and hills. Only in a certain area has inward occurred. Looking at the general shape of the surface of Y2A, the number of peaks and pits is quite high, although the depth of peaks is not very high. This supports the view that the Y2A specimen will not be inferior in terms of adhesion, which is also deduced from other figures. The excess number of peaks and troughs will play an important role in the effect of the plates against each other.

When the specimen with paper surface in Figure 2.12 is examined, if the 100 μm non-wear surface is accepted as the reference limit, the specimen with Y2 Figure 2.12 (a) is worn very little and homogeneously. A layer of approximately 25 μm is separated from the surface. If we refer to the region shown as very deep and blue in Figure 2.12(a), the beam sent by the optical profilometer device to measure from the surface is caused by contact with an area outside the specimen, so it is not taken into account. Except for the neglected area, the wear depth of the specimen with Y2 is partially reduced to 50 μm, but there is a homogeneous distribution in general. If the O Figure 2.12 (b) is examined, the trace of linear wear is clearly visible, and the wear depth decreased from approximately 100 μm to 10 μm. Microscope and post-wear images (Figure 2.8 and Figure 2.9d) support the wear occurring on the O. In this respect, it can be said that the specimen with Y2 is more resistant to abrasion.

When evaluated in terms of roughness, the roughness value of the O specimen may be too high, especially Rz (maximum peak to trough difference values), but the number of peaks and troughs in the O specimen Figure 2.12 (b) appears to be quite low, so the O specimen has a low roughness value. Consequently, the O sample is predicted to have low adhesion ability. When looking at the Y specimen Figure 2.12 (a), the surface showed a fairly uniform wear. Not too many peaks and troughs are formed and therefore the Rz value will be low. However, the Ra (average surface roughness value) value is higher compared to O. Consequently, it is possible to say that the adhesion ability of the specimen with Y2 will be higher. It should not be forgotten here; these topography images were measured over 2 mm × 2 mm areas. Therefore, the images do not represent the entire surface or all worn sections. The comments and discussion are obtained by evaluating the microscope (Figure 2.9), macro-scale (Figure 2.10), and post-test specimen images (Figure 2.8) together with friction coefficient results (Figure 2.6 and Figure 2.7).

2.3.4 Result of measurement torque transmission

When the clutch mechanism is not active, the movement and torque from the engine is transmitted to the compressor via friction plates. As can be seen in Figure 2.1, the friction plates that receive the movement from the motor and the friction plates connected to the compressor are in surface-to-surface contact with each other, and the frictional effect on the surfaces causes the plates on the motor side to rotate plates on the compressor side and torque transmission occurs. Here, as the plates force each other to rotate, the frictional effect produces an effect in the opposite direction to the direction of rotation. This effect that forces the rotation is due to the torque generated during friction. Moreover, the friction torque that makes rotation difficult also varies depending on operating speeds.

Experiments were carried out on the Y2 specimen to find these torque transmission values that occur depending on the operating speeds between plates. The transmission torque values are shown in Figure 2.13.

It can be seen from Figure 2.13 that there is an inverse proportion between operating speed and torque values. The transmitted torque value increased as the operating speed decreased. At low speeds, the adhesion

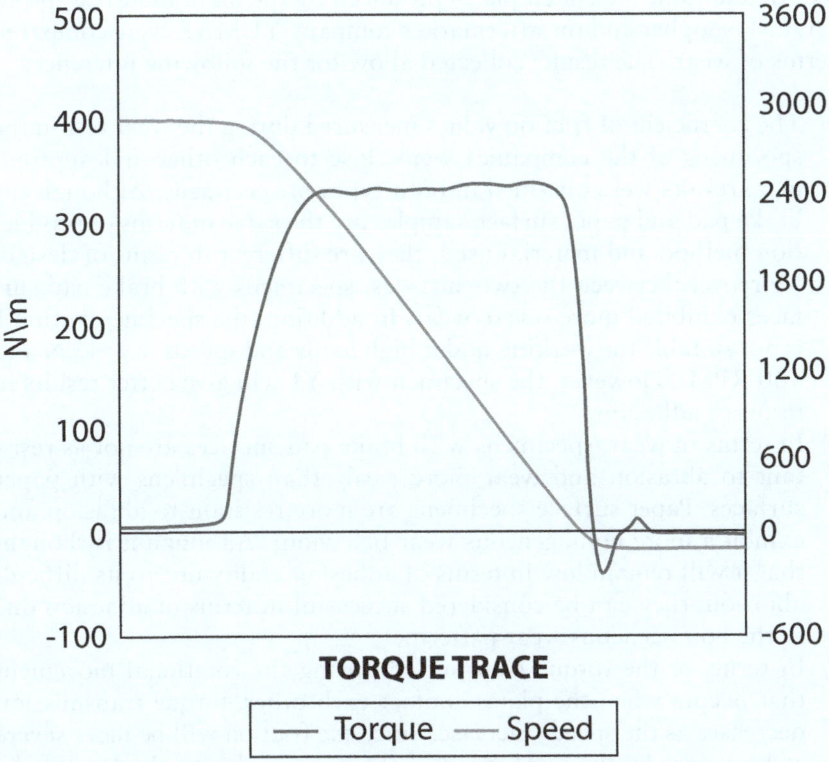

Figure 2.13 Torque Transmission graph for Y2.

between the interfaces will be higher because the surfaces are in better contact with each other. Therefore, the torque received from the engine will be transmitted to the compressor with small loss. However, when speeds increase, the friction between the surfaces will increase and therefore, the surfaces will wear out and will not be in good contact with each other. As a result of the wear test, the wear of the plates can be understood from Figures 2.8–2.12. Consequently, the torque transmission will approach stalling at high speeds.

2.4 CONCLUSIONS

The wear caused by the contact of the friction plates in the multi-disc clutch mechanism was investigated. Wear experiments were carried out using a pin-on-disc tribometer at two different loads and speeds to explain the friction caused by the contact of the plates with each other. The friction coefficient resulting from wear and surface images after wear were investigated. In addition, the clutch plates produced by the main industrial product (OEM) supplier and the aftermarket company YUMAK were compared in terms of wear. The results collected allow for the following inferences:

(i) The coefficient of friction values measured during the wear test in the specimens of the companies were close to each other and approximate results were obtained in both types of specimens. Although the brake pad and paper surface samples are the same in terms of production method and material used, they are different in terms of design. Moreover, between the two surfaces, specimens with brake pad surfaces exhibited more severe wear. In addition, the specimen with Y1 is not suitable for working under high loads and speeds (e.g. 45 N and 400 RPM). However, the specimen with Y1 will give better results in terms of adhesion.

(ii) In terms of wear, specimens with brake pad surfaces are not as resistant to abrasion and wear more easily than specimens with paper surfaces. Paper surface specimens are more resistant to abrasion and exhibit a more homogeneous wear behaviour. Although it is thought that it will remain low in terms of adhesion ability due to its difficult abrasion, they can be considered successful in terms of adhesion due to the homogeneous wear pattern.

(iii) In terms of the torque transmitted during the rotational movement that occurs when the plates contact each other, torque transmission decreases as the speeds increase. Since the friction will be more severe at high speeds, the weakening of the contact due to the loss of the interface where the transmission is made reduces the transmitted torque values.

REFERENCES

1. Budynas, F.C. and J.K. Nisbett, *Shigley's Mechanical Engineering Design*. New York: McGraw-Hill Education, 2015: p. 574–621.
2. Lai, Y.G., Simulation of heat-transfer characteristics of wet clutch engagement processes. *Numerical Heat Transfer, Part A Applications*, 1998. 33(6): p. 583–597.
3. Yang, Y., R.C. Lam, and T. Fujii, Prediction of torque response during the engagement of wet friction clutch. *SAE Transactions*, 1998:107 p. 1625–1635.
4. Biczó, R., et al., Composite friction materials for brakes and clutches. *Scientific Bulletin Series C: Fascicle Mechanics, Tribology, Machine Manufacturing Technology*, 2016. 30: p. 21.
5. Ingram, M., et al., Contact properties of a wet clutch friction material. *Tribology International*, 2010. 43(4): p. 815–821.
6. Scott, W. and P. Suntiwattana, Effect of oil additives on the performance of a wet friction clutch material. *Wear*, 1995. 181: p. 850–855.
7. Zagrodzki, P. and S.A. Truncone, Generation of hot spots in a wet multidisk clutch during short-term engagement. *Wear*, 2003. 254(5–6): p. 474–491.
8. Zagrodzki, P. and P. Zagrodski, Influence of design and material factors on thermal stresses in multiple disc wet clutches and brakes. *SAE Transactions*, 1991: p. 395–405.
9. Nyman, P., et al., Influence of surface topography on friction characteristics in wet clutch applications. *Wear*, 2006. 261(1): p. 46–52.
10. Yu, L., et al., Investigation on the failure mechanism and safety mechanical-thermal boundary of a multi-disc clutch. *Engineering Failure Analysis*, 2019. 103: p. 319–334.
11. Iqbal, S., et al., Modeling and analysis of wet friction clutch engagement dynamics. *Mechanical Systems and Signal Processing*, 2015. 60: p. 420–436.
12. Natsumeda, S. and T. Miyoshi, (April 1, 1994). Numerical simulation of engagement of paper-based wet clutch facing. ASME. J. Tribol. April 1994; 116(2): 232–237.
13. Yu, L., et al., Study on the groove area effects on the friction-wear properties of a Cu-based wet clutch. *Tribology International*, 2021. 163: p. 107096.
14. Mansouri, M., et al., Thermal and dynamic characterization of wet clutch engagement with provision for drive torque. *Journal of Tribology*, 2001. 123(2): p. 313–323.
15. Bąk, M., Torque capacity of multidisc wet clutch with reference to friction occurrence on its spline connections. *Scientific Reports*, 2021. 11(1): p. 21305.
16. Marklund, P. and R. Larsson, Wet clutch friction characteristics obtained from simplified pin on disc test. *Tribology International*, 2008. 41(9–10): p. 824–830.
17. Fei, J., et al., (August 12, 2008). Carbon-fiber reinforced paper-based friction material: Study on friction stability as a function of operating variables. ASME. J. Tribol. 2008;130(4): 041605.
18. Vaidyaraman, S., et al., (2001). C/SiC material evaluation for aircraft brake applications. In Krenkel, W., Naslain, R., & Schneider, H., (Eds.), *High Temperature Ceramic Matrix Composites*, 2001: p. 802–808.

19. Marklund, P., Permeability measurements of sintered and paper based friction materials for wet clutches and brakes. *SAE International Journal of Fuels and Lubricants*, 2010. 3(2): p. 857–864.
20. Kim, S.H. and H.S. Lee, Effect of pressure on tribological characteristics between sintered friction materials and steel disk. *International Journal of Precision Engineering and Manufacturing*, 2011. 12: p. 643–650.

Chapter 3

Prediction and assessment of surface roughness in wire-arc-based directed energy deposition process

Yoshit Tiwari, Diya Mukherjee, Manidipto Mukherjee, and Soumyajit Kundu

3.1 INTRODUCTION

The wire-arc-based directed energy deposition (WADED) technique, also known as wire arc additive manufacturing (WAAM), has emerged as a promising metal additive manufacturing technology for the fabrication of medium and large-scale metal parts [1]. In the WADED process, metal wire is melted and deposited layer by layer using an electric arc. This process incorporates welding power sources such as gas metal arc welding (GMAW), gas tungsten arc welding (GTAW), and plasma arc welding (PAW) to serve as a heat source (welding arc). Additionally, it employs industrial robotic arms or CNC machines to achieve precise position and motion control, supported by appropriate software and hardware integration. The fundamental components of the WADED system are depicted in Figure 3.1a. WADED presents an economical approach to additive manufacturing due to its ability to utilize less costly wire feedstock than powders or other raw material forms [2]. The technique exhibits high deposition rates, facilitating the rapid production of large-scale components, making it particularly suitable for time-critical applications. Moreover, WADED allows for using various metallic materials, such as aluminum, titanium, stainless steel, and nickel alloys, providing the flexibility to produce components with diverse mechanical and physical properties [3].

Being a scalable process, WADED can be tailored to manufacture components of different sizes, ranging from small intricate parts to large structures, making it well-suited for numerous industrial applications. It can potentially revolutionize production processes in aerospace, automotive, oil and gas, nuclear, shipbuilding, mining, and heavy equipment, primarily due to its cost-effectiveness in equipment and raw materials and ease of handling and operation. Comparative studies have shown that WADED exhibits higher efficiency in manufacturing large-scale, medium- to low-complexity metal parts compared to other metal additive manufacturing processes like laser powder directed energy deposition and powder bed fusion [4]. However, specific challenges related to surface roughness,

DOI: 10.1201/9781003713364-3

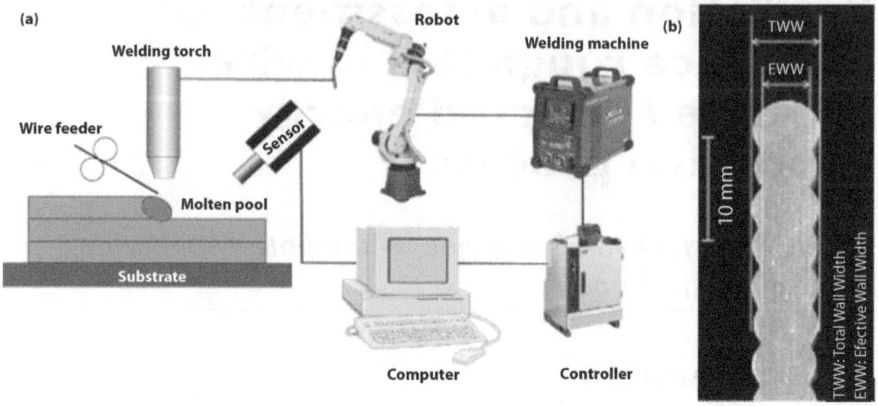

Figure 3.1 (a) Different components of the WADED system [8] (Yan Li et al., Elsevier B.V, 2021, CC-BY-4.0) and (b) the cross-sectional view of WADED-processed wall structure [9] (Silva et al., MDPI, 2021, CC-BY-4.0).

dimensional accuracy, process monitoring, design considerations, and post-processing requirements must be carefully addressed to fully exploit the potential of the WADED process [5–7]. Overcoming these challenges would enable WADED to revolutionize manufacturing processes by enabling the production of complex and functional metal components with improved efficiency and reduced costs.

As the WADED process involves a sequential deposition of layers (Figure 3.1a), every layer's surface characteristics directly influence the final components' overall quality. Figure 3.1b presents a cross-sectional view of a WADED-processed wall structure, showing that the deposited beads overlap with irregular and undulating side surfaces. The surface roughness of each layer is primarily governed by the bead geometry, which exhibits significant variations depending on the chosen process parameters, deposition path, and strategies. Inadequate surface quality in any layer can accumulate throughout the build, resulting in diminished overall accuracy of the final component. Moreover, substandard surface quality can lead to non-uniform deposition in subsequent layers, introducing potential defects such as porosity, voids, and delamination [9]. These factors have a detrimental impact on the functional properties of the components, including the strength of industrial parts and the fatigue life of manufactured elements [10]. The post-processing efforts to address poor surface quality can be labour-intensive and time-consuming, mainly due to the intricate shapes of the deposited components. This manual post-processing approach may also result in substantial wastage of both manual labor and feedstock materials [6]. Consequently, implementing the WAAM process for industrial production may lose its advantageous appeal. Therefore, achieving high surface quality is of utmost significance in WAAM production. Ensuring proper surface characteristics throughout the deposition process is crucial to producing components with enhanced performance and mechanical properties, thereby fostering the successful adoption of WAAM in industrial applications.

Controlling surface roughness is a critical factor in optimizing the quality of the WADED process and reducing the need for additional post-processing expenses. Unlike the traditional understanding of surface roughness in conventional machining, surface roughness in additively manufactured structures refers to the sidewall formed by stacked layers and the waviness of the top surface resulting from overlapping beads. Ding et al. [11] have highlighted that the WADED process faces significant challenges related to poor accuracy and surface finish, consequently hindering its broader industrial application.

Several factors significantly influence surface roughness in WADED processes, including the geometry of individual beads, the degree of bead overlap, and the temperatures of the substrate and interlayer. Additionally, distortion and shrinkage, as well as the material's properties, play crucial roles. Controlling the process parameters, such as current, voltage, wire feed speed (WFS), welding speed (WS), contact tip to workpiece distance (CTWD), and substrate temperature, allows for the regulation of bead geometry. Through the optimization of these parameters, it becomes possible to achieve a continuous and smooth bead profile, leading to a reduction in both side and top surface roughness of the deposited structure. While a smooth bead profile aids in minimizing surface roughness, the extent of bead overlap significantly impacts the top surface waviness of the deposited structure. Determining the appropriate amount of bead overlap depends on the type of substrate surface (flat or inclined). It is influenced by the characteristics of the bead profile, such as bead width and height.

3.2 DEFINITION OF SURFACE ROUGHNESS

Generally, surface roughness pertains to the existence of orderly or disordered irregularities on a designated surface. The surface profile of an item displays a multifaceted arrangement encompassing peaks and valleys characterized by differing elevations, depths, and separations. In the context of the WADED methodology, the analysis of surface roughness can be categorized into two distinct classifications: the appraisal of upper surface roughness (as illustrated in Figure 3.2a,b) and the evaluation of lateral surface roughness (as depicted in Figure 3.2a,c). The top surface roughness or waviness primarily depends on bead offset or overlap factors. On the other hand, the side surface roughness is predominantly influenced by welding process parameters, including welding speed, wire feed speed, current, voltage, CTWD, deposition rate, and the material used. Additionally, specific other parameters such as depositing strategies (e.g., interlayer temperature,

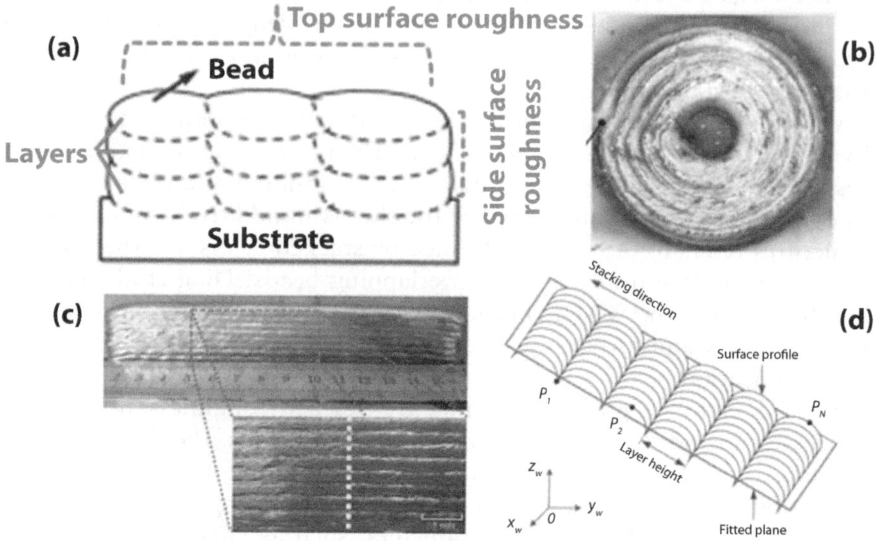

Figure 3.2 (a) Schematic representation of top and side surface roughness. (b) Top surface of deposited structure [15] (Biester et al., Elsevier B.V, 2022, CC BY-NC-ND-4.0). (c) Side surface roughness of thin wall deposited via WADED process, indicating a stair-stepping effect. (d) Schematic of surface roughness profile with fitted plane [13] (Xiong et al., Springer Open, 2018, CC-BY-4.0).

deposition path strategies, forced heating, and cooling systems) can also impact the surface roughness of the fabricated parts.

Numerous scientific investigations have been undertaken to accurately define surface roughness within the WADED (wire-arc-based directed energy deposition) process framework. Yehorov [12] introduced a methodology aimed at quantifying surface waviness through the computation of the effective area (EA) ratio derived from a cross-sectional analysis performed orthogonally to a wall section. The EA ratio is ascertained by dividing the total cross-sectional area of the wall by the area of the largest inscribed rectangle contained within it. Nevertheless, this approach is constrained by its reliance on a solitary cross-sectional assessment, posing a challenge in comprehensively capturing the holistic surface roughness attributes across the entire wall structure. To surmount this limitation, researchers have pursued alternative strategies. As mentioned earlier, Xiong et al. [13, 14] assessed surface roughness by quantifying the spatial deviations between individual surface profile points and the mean plane established by the surface. Figure 3.2c portrays a typical surface profile of a fabricated component using the WADED technique, revealing a conspicuous non-uniform surface texture, particularly at contiguous strata. This phenomenon, recognized as the “stair-stepping” phenomenon, is a prevalent inherent attribute within additive manufacturing methodologies. It arises when the layer height surpasses 1 mm during the deposition of multiple layers. Notably, the stair-stepping effect is more prominently manifested in the WADED process than in powder-based DED and powder bed fusion techniques. This discrepancy arises from utilizing a heightened layering dimension in the WADED process. Ultimately, the surface integrity of deposited components significantly governs the extent of material machining requisite in subsequent refining procedures.

Utilizing the surface profile provided in Figure 3.2c, a schematic diagram, as depicted in Figure 3.2d, is developed to visually convey the surface roughness of components produced through the WADED process. Within this context, the points (P_1, P_2, …, P_N) are assumed to be positioned on the surface profile within a global coordinate system. Considering the inherent randomness in multilayer depositions, the least squares method calculates an approximate plane. The mathematical representation of the fitted plane can be described by the following expression (Equation 3.1) [12], where *A*, *B*, *C*, and *D* are the coefficients of the fitting planes, respectively, and *x*, *y*, and *z* represent the variables within the world coordinate system.

$$Ax + By + Cz + D = 0 \tag{3.1}$$

The mathematical representation (Equation 3.2) [12] delineates the separations of individual surface profile points from the adjusted plane. *N* signifies the aggregate count of points used to fit in this equation.

$$\sum d_i = \sum_{i=1}^{N} \frac{|Ax_i + By_i + Cz_i + D|}{\sqrt{(A^2 + B^2 + C^2)}} \tag{3.2}$$

Then, the average surface roughness factor (*R*) is defined by the following expression [12]:

$$R = \frac{\sum d_i}{N} \tag{3.3}$$

3.3 MEASUREMENT OF SURFACE ROUGHNESS IN THE WADED PROCESS

The simplest way to measure the surface roughness of the wall structure deposited by WADED processes is to cut the cross-section of the wall and measure the total wall width (TWW) and effective wall width (EWW) by capturing a 2D image of the cross-section (Figure 3.1b) and calculating surface roughness/waviness (*R*) using the following expression [9, 12]:

$$R = \frac{TWW - EWW}{2} \tag{3.4}$$

This method utilizes a destructive approach to calculate surface roughness, yielding only an average quantitative value of R. Consequently, it is unsuitable for in situ prediction and modeling of surface roughness. To address this limitation, alternative non-destructive techniques, contact or non-contact, can be employed to assess the surface roughness of WADED-processed surfaces. A prevalent technique utilized in assessing surface roughness entails using a stylus profilometer through direct contact. Nevertheless, this methodology is characterized by its protracted temporal demands and labor-intensive nature, rendering it less congruent with the needs of the WADED process. The WADED process pertains explicitly to the fabrication of extensive structural configurations necessitating expedited measurement procedures. In contradistinction, optical measurements based on laser vision offer a constellation of benefits, encompassing non-contact modality, swift data acquisition, and consistent findings. As a result, we strongly advocate for the integration of laser vision-based measurement methodologies within the framework of the WADED process. Notably, a cohort of investigators has effectively harnessed laser vision systems for this precise intention [13–16].

Figure 3.3 illustrates the configuration of a laser vision system capable of functioning as either a 3D scanner or a 2D profiler. For instance, Chen et al. [17] employed a 3D laser scanning system to quantify the roughness of the top surface. Once the scanning process is complete, point cloud data is acquired, necessitating subsequent processing to extract the roughness

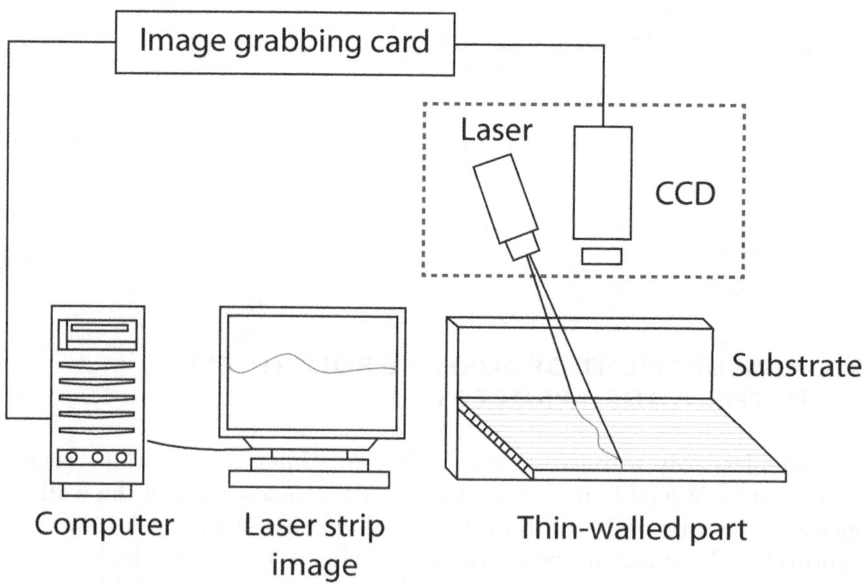

Figure 3.3 Schematic diagram representing surface roughness profile measurement using a laser vision system [13] (Xiong et al., Springer Open, 2018, CC-BY-4.0).

information from the top surface. Due to the laser scanning system's ability to capture a larger area than the actual top surface of the fabricated component, it becomes essential to filter out the surplus point cloud data. To achieve this, the authors utilized multiple planes fitted to the point cloud data to eliminate the excessive data points. The top surface roughness was accurately determined by considering the data points above the resultant plane. Similarly, Xiong et al. [14] employed a laser scanner to evaluate the roughness of side surfaces. They collected data sets from which a plane was fitted using Equations 3.1–3.3, and the surface roughness was calculated as the variance of the data sets relative to this fitted plane.

3.4 PROCESS PARAMETERS RELATED TO SURFACE ROUGHNESS

Various process parameters can significantly influence the top and side surface roughness of structures fabricated through WADED. Numerous research investigations have been dedicated to establishing the correlations between the control variables in the WADED process and the resulting surface roughness. Zhang et al. [18] observed that the surface finish of Inconel 625 deteriorated with increasing heat input during the deposition process. Additionally, they found that a higher heat input reduced the number of layers required to achieve a specific wall height. Jiang et al. [19] reported that increased WFS led to an elevated average surface roughness value in the deposited structures when heat input was also increased. Le et al. [20] deposited stainless steel using various heat inputs and concluded that higher heat inputs negatively affected the surface quality of the deposited wall structures during the WADED process. Xiong et al. [14] investigated the effects of process parameters on surface roughness of multilayer single-pass wall structures fabricated through GMAW-based WADED. They found that increasing WFS may lead to higher surface roughness and reduced layer height, while increasing travel speed may simultaneously decrease surface roughness and layer height. Dinovitzer et al. [21] investigated the impact of travel speed (TS) and electric current (I) on the characteristics of surface roughness. Their analysis revealed that heightened TS and reduced I values correlate with a degradation in surface roughness quality. In an independent study, Teixeira et al. [22] explored the cumulative influence of TS and interlayer temperature on the deposition process of low-carbon, low-alloy steel. The study findings disclosed that an escalation in TS yields a noteworthy escalation in surface roughness, whereas the influence of interlayer temperature exhibited relatively subdued effects. Employing the WADED technique, Xiong et al. [23] fabricated wall structures utilizing 300M steel, with variations in the interlayer temperature spanning 100°C to 600°C. Their observations established that an interlayer temperature of

200°C resulted in the most minimal degree of surface waviness. This underscores a distinctive V-curve correlation between surface waviness and interlayer temperature.

Several researchers have explored incorporating an extra-active cooling system during deposition to address the challenges of wall widening and surface roughness during the WADED process. This approach aims to control wall widening caused by excess heat accumulation and reduce heat transfer rates as height increases. Dash et al. [24] introduced a heat exchanger integrated into the substrate to enhance active cooling during deposition. They observed that the implementation of active cooling led to a remarkable 99% reduction in volumetric/surface defects. Similarly, Le et al. [25, 26] employed additional dry air cooling to enhance the internal and external quality of SS308L deposited parts. They evaluated the side surface roughness of the fabricated wall using a 3D scanner with and without forced air cooling. The study conducted by Le et al. demonstrated that increasing the interlayer wait time from 30 seconds to 60 seconds resulted in a marginal improvement, reducing the average surface roughness value from 0.24 mm to 0.22 mm. However, with forced dry air cooling, even a wait time of 10 seconds led to a superior surface finish, with Ra values of 0.17 mm. Scotti et al. [27] and Long et al. [28] employed near immersion active cooling (NIAC) in their research, submerging the substrate and depositing the wall partially in a liquid coolant (water was utilized in both cases). The NIAC approach facilitated rapid cooling of the deposited layer, thereby mitigating heat accumulation and reducing wall widening caused by excessive flow of molten metal during the deposition process. The findings revealed that NIAC resulted in a smoother wall surface with improved surface roughness compared to the absence of NIAC. Previous investigations have extensively explored the impact of various intrinsic and extrinsic process parameters on surface roughness. A comprehensive overview of these influencing process parameters is presented collectively in Figure 3.4.

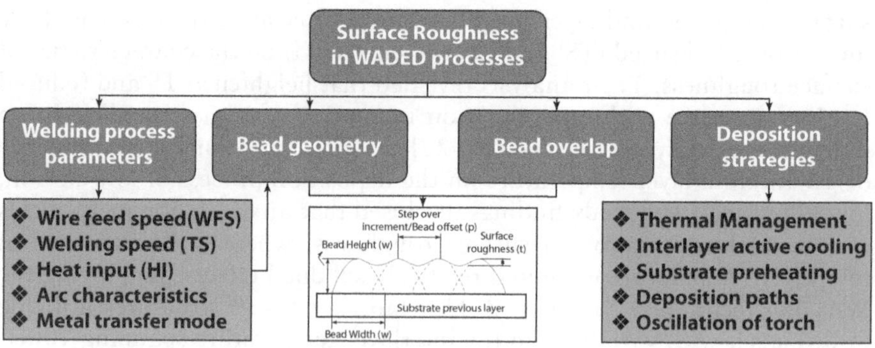

Figure 3.4 Parameters influencing surface roughness in the WADED process.

3.5 BEAD MODELING

3.5.1 Single-bead modeling

Typically, investigators employ three frequently utilized mathematical functions (namely the parabola, cosine, and arc functions) to depict the cross-sectional configuration of an individual weld bead [29–32], as outlined in Table 3.1. The dimensions of deposition height (h), deposition width (w), and area (A) for each specific model are ascertained using distinct model parameters. These functions incorporate model parameters, designated as a, b, and c, necessitating determination via empirical assessments.

Ding et al. [33] utilized a trifold approach in their investigation, incorporating the full spectrum of the three models. Their findings advocate that, within the empirical models scrutinized, the parabolic and cosine models present superior precision in approximations when compared with the arc model. The parabolic model exhibits remarkable concurrence with empirical outcomes across diverse deposition speeds. Additionally, the researchers emphasized the significance of incorporating the parabolic bead profile into the framework for modeling multi-bead overlap, a crucial step to achieve enhanced fidelity in predicting top surface roughness.

3.5.2 Multi-bead overlap models

The separation distance (denoted as "d" in Figure 3.5) between neighboring beads is pivotal in determining surface quality and texture. When this separation distance (d) surpasses an individual bead's width (w), an absence of overlap occurs among adjacent beads. It can be inferred that under such circumstances, the undulations on the upper surface are nearly equivalent to a single bead's height (h). Nevertheless, with a reduction in the separation distance, the extent of overlapping region, as illustrated in Figure 3.5a, becomes more pronounced, simultaneously leading to a reduction in the valley region. As the separation distance decreases, a critical juncture is reached wherein the overlapping region aligns with the valley region, culminating in an optimally level surface. However, upon further reduction of the separation distance, an excessive overlap manifests, resulting in an escalation of the layer thickness and a diminishment of surface evenness. Thus, the ideal separation distance (d), bead offset (p), or step-over increment in overlap models is ascertained based on the principle that a level surface is attained when the extent of overlap matches that of the valley region. The conceptual framework developed on this premise is the flat-top overlapping model (FOM), as depicted in Figure 3.5b [29,34]. In the conventional model of overlapping weld beads with a flat-top profile, as shown in Figure 3.5b, the valley region is defined by the confines of two adjacent beads and a straight line connecting the highest points of both weld beads. Nonetheless, as documented in earlier scholarly works, prior

Table 3.1 Three frequently used bead models related to bead profile

Model	*Model function*	*Height, h*	*Width, w*	*Area*	*Reference*
Parabola model	$y = ax^2 + c$	c	$2\sqrt{-\frac{c}{a}}$	$A_p = \frac{4c}{3}\sqrt{-\frac{c}{a}}$	[32]
Cosine model	$y = a\cos(bx)$	a	$\frac{\pi}{b}$	$A_c = \frac{2a}{b}$	[32]
Arc model	$y = \sqrt{a^2 + x^2} + b$	a–b	$2\sqrt{a^2 - b^2}$	$A_a = \arccos \arccos\left(-\frac{b}{a}\right) - b\sqrt{a^2 - b^2}$	[32]

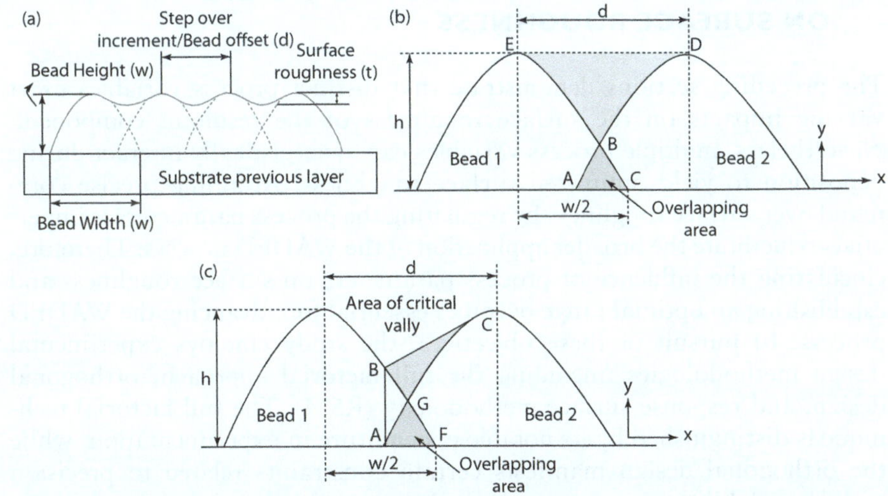

Figure 3.5 Schematic representation of the overlapping bead models (a) overlapping beads and top surface roughness, (b) flat-top overlapping model (FOM), and (c) tangent overlapping model (TOM) [35] (Banaee et al., Taylor & Francis Group, 2023, CC-BY-NC-4.0).

investigations into welding processes revealed the presence of a distinctive geometry referred to as the "critical valley." This critical valley configuration offers a more precise depiction of the cross-sectional characteristics, which are especially relevant when dealing with multiple welding beads in the WADED process. To surmount this limitation inherent in the FOM, Ding et al. [33] introduced an alternative approach known as the tangent overlapping model (TOM). The fundamental disparity between the FOM and the TOM resides in their respective conceptualizations of the formation of the uppermost surface. In the context of the FOM, a pivotal offset distance (d_{cr}) of 0.66 times the bead width (w) is ascertained, operating under the assumption that a planar surface is established when the areas of overlap and the valley are equivalent (for instance, Area (ABC) = Area (BED) as depicted in Figure 3.5a). Conversely, the TOM postulates that the upper surface is tangential to the peak of the subsequently deposited second bead. This identifies a critical valley configuration, visually represented in Figure 3.5c. Through the equating of volumes encompassing the critical valley and the regions of overlap (BCG and AGF, respectively, as shown in Figure 3.5c), a revised critical offset distance (d_{cr}) of 0.738 times the bead width (w) is derived. It is important to emphasize that the aforementioned critical offset distance is contextually specific to weld beads exhibiting a parabolic cross-sectional profile, a characteristic commonly observed in WADED processes.

3.5 OPTIMIZATION OF PROCESS PARAMETERS ON SURFACE ROUGHNESS

The preceding sections demonstrate that distinct process variables exert varying impacts on the surface roughness of the resultant component. Nevertheless, multiple process variables can synergistically interact during deposition to yield optimized surface roughness. Attaining precise command over surface roughness by regulating the process parameters is imperative to facilitate the broader application of the WADED process. Therefore, elucidating the influence of process parameters on surface roughness and establishing an optimal range becomes essential for advancing the WADED process. In pursuit of these objectives, the study employs experimental design methodologies, including the full factorial approach, orthogonal design, and response surface methodology (RSM). The full factorial technique is distinguished by its notable expenditure in experimentation, while the orthogonal design manifests certain constraints related to precision and dependability. Conversely, the RSM garners distinction due to its economical nature and enhanced dependability, establishing it as the favored avenue for this investigation. Zhang et al. [36] conducted a comprehensive investigation into the influence of specific operational parameters on the surface roughness of components fabricated through the WADED process. The parameters under scrutiny encompassed WFS, TS, and the cold metal transfer/pulse (CMT/P) ratio. A systematic approach rooted in statistical techniques, including the design of experiments and analysis of variance (ANOVA), was employed to enhance these process variables and construct precise regression models systematically. In addition to these methodologies, a refined non-dominated sorting genetic algorithm (NSGA-II) was utilized for parameter optimization. The overarching goal of this optimization was to minimize the degree of surface roughness in the resultant additive components. Subsequently, the researchers successfully established a compelling association between the surface roughness of the fabricated parts and the aforementioned critical process variables, which is concisely represented in Equation 3.5, where $X1$, $X2$, and $X3$ are WFS, TS, and CMT/P ratio, respectively.

$$\begin{aligned} Surface\,Roughness = {} & 199.11 - 71.98X1 - 7.8X2 - 10.45X3 - 2.38X1 \\ & \times X2 + 0.91X2 \times X3 + 11.8X1^2 + 0.87X2^2 + 0.66X3^2 \end{aligned} \tag{3.5}$$

The study established a significant correlation between WFS and TS with surface roughness. Elevating the WFS was associated with a conspicuous reduction in surface roughness. Conversely, the influence of travel speed on surface roughness exhibited an initial decline followed by an increase as travel speed escalated. Employing a ratio coefficient of 2 led to an elevated level of formation quality. Employing optimization techniques, the investigation identified 13.73 mm as the minimal achievable surface roughness, signifying the utmost optimal outcome.

3.6 MODELING AND PREDICTION OF SURFACE ROUGHNESS USING MACHINE LEARNING

In the WADED technique, which involves the incremental fabrication of components, the ultimate quality of the end product is substantially impacted by the surface characteristics exhibited by each layer. Throughout the WADED procedure, the successive deposition of material beads forms a cohesive layer through bead overlap. The topographical uniformity of these layers is fundamentally governed by factors such as the geometric attributes of the individual beads and the extent of overlap between adjacent beads. These parameters, subject to variation based on distinct process variables and path planning methodologies, play a pivotal role in determining the surface flatness of each stratum. Efforts to refine surface attributes in the context of WADED have progressively embraced the integration of machine learning algorithms (Figure 3.6). These algorithms have assumed a central role in the comprehensive analysis of voluminous datasets generated during the additive deposition process. Moreover, combining machine learning techniques with the WADED process has recently gained remarkable

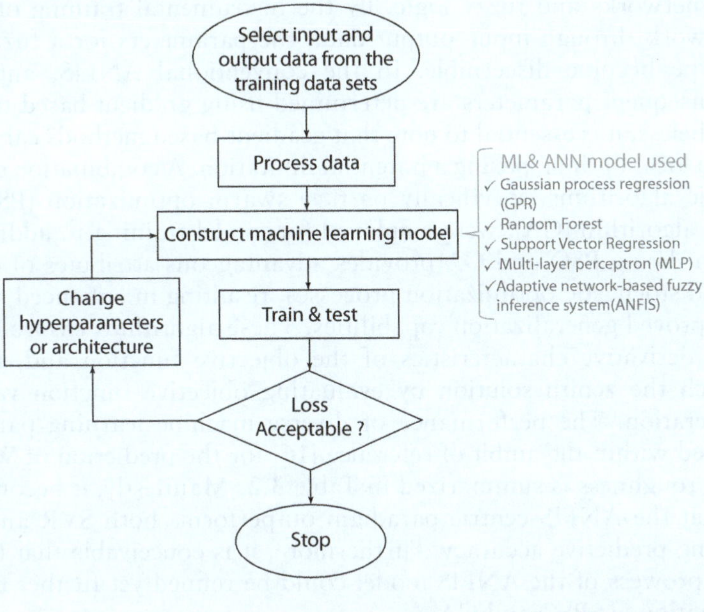

Figure 3.6 Flow diagram indicating construction procedure of applied machine learning models for surface roughness modeling and prediction in the WADED process.

prominence. This section comprehensively assesses diverse machine learning tools harnessed to facilitate the fabricated components' vertical and lateral surface roughness.

In the investigation by Xia et al. [16], the anticipation of surface roughness during the WADED process was executed by leveraging the influence of process parameters. This research encompassed an assortment of machine learning methodologies for the prognosis of surface roughness, notably encompassing the adaptive neuro-fuzzy inference system (ANFIS), support vector regression (SVR), and extreme learning machine (ELM). ANFIS, SVR, and ELM represent three prominent exemplars of machine learning paradigms, each founded upon distinctive principles. ANFIS amalgamates linguistic tenets of fuzzy logic with the adeptness of neural networks in addressing regression challenges. Significantly, the ANFIS paradigm offers celeritous learning and adaptability as its chief virtues. Conversely, SVR, a comparatively recent entrant in machine learning, generates impressive regression performance by minimizing an upper bound for the cost function. Its outperformance of erstwhile algorithms has substantiated this. ELM, heralded as a vanguard technique for training neural networks, is notably recognized for its expeditious training of neural networks in the multilayer perceptron (MLP) genre [33]. These paradigms offer several key advantages, encompassing (i) independence from explicit mathematical formulations for intricate processes, (ii) facile implementation, and (iii) robust fitting and generalization capabilities. Furthermore, ANFIS emerges as an adaptive fuzzy logic inference system that amalgamates the strengths of neural networks and fuzzy logic. By the instrumental training of a neural network through input–output data, the parameters for a fuzzy logic archetype become discernible. In the conventional ANFIS, antecedent and consequent parameters are determined using gradient-based training. Nevertheless, it is essential to note that gradient-based methods can be sensitive to local optima, posing a potential limitation. A combination of metaheuristic algorithms, specifically particle swarm optimization (PSO) and genetic algorithm (GA), were employed for model training to address this concern. Using PSO and GA provides advantageous attributes of concurrent and stochastic optimization processes, resulting in enhanced stability and improved generalization capabilities. These algorithms ensure reliance on the derivative characteristics of the objective function and, instead, approach the zenith solution by evaluating objective function values at each iteration. The performance of diverse machine learning paradigms employed within the ambit of reference [16] for the prediction of WADED surface roughness is summarized in Table 3.2. Manifestly, it becomes evident that the ANFIS-centric paradigm outperforms both SVR and ELM regarding predictive accuracy. Furthermore, it is conceivable that the predictive prowess of the ANFIS model could be refined yet further through optimization via PSO and GA.

Table 3.2 Comparison of performance characteristics of ML models

Model	*RMSE*	R^2	*MAE*	*MAPE (%)*	*Max. dev.*	*Min. dev.*
ANFIS	0.0802	0.9189	0.0688	16.93	0.1751	0.0118
GA–ANFIS	0.0694	0.93516	0.0574	14.15	0.1516	0.0033
PSO–ANFIS	0.0757	0.9225	0.0629	15.65	0.1758	0.0006
ELM	0.0985	0.88691	0.0761	21.33	0.2093	0.0005
SVR	0.1005	0.862	0.0719	17.21	0.2261	0.0001

Source: Xia et al., J. Intell. Manuf. 33 (2022) 1467–1482.

Chen et al. [17] employed the Gaussian process regression (GPR) machine learning technique to model and forecast the surface roughness of a material's top layer. In their investigation, they manipulated input parameters, namely, weaving frequency shift (WFS), thread spacing (TS), weaving amplitude (W_a), and weaving wavelength (W_w), within the context of 45 distinct experimental trials. Among these, 45 datasets were designated for training the predictive model, while five additional datasets were reserved for model validation. A 3D laser scanning method was employed during the training phase of the GPR model to quantify the resultant surface roughness. The results have proven effective, demonstrated by a maximal predictive deviation of under 50 μm. This attainment serves to highlight the precision and dependability of this method. Conversely, an alternative approach involving a MLP-based machine learning technique was applied by researchers in predicting surface roughness for a woven path-deposited structure in WADED [37], as shown in Figure 3.7. The input parameters consisted of WFS, TS, W_a, and W_w, while the predicted output was the surface roughness itself. The ReLU (rectified linear unit) function served as the activation function within the hidden layer, and mean squared error was utilized as the evaluation metric for the model's performance. The study identified an optimal MLP architecture with a minimal mean squared error value of 0.00052. This observation highlights the potential of the MLP technique for accurate surface roughness prediction, even when working with a reduced dataset.

Yaseer and Chen [38] conducted a comparative study involving the utilization of random forest and MLP techniques for the prediction of surface roughness in the context of the WADED process (Figure 3.7). The random forest algorithm, renowned for its adeptness in capturing intricate nonlinear phenomena, was employed due to its applicability in modeling complex processes. The efficacy of the random forest model was evaluated and juxtaposed against that of the MLP model concerning the estimation of layer roughness. This evaluation utilized specific input parameters: travel speed, wire feed speed, weaving amplitude, and wavelength. Empirical trials were executed to meticulously examine the competence of the random forest and MLP models in elucidating the intricate interplay between parameters

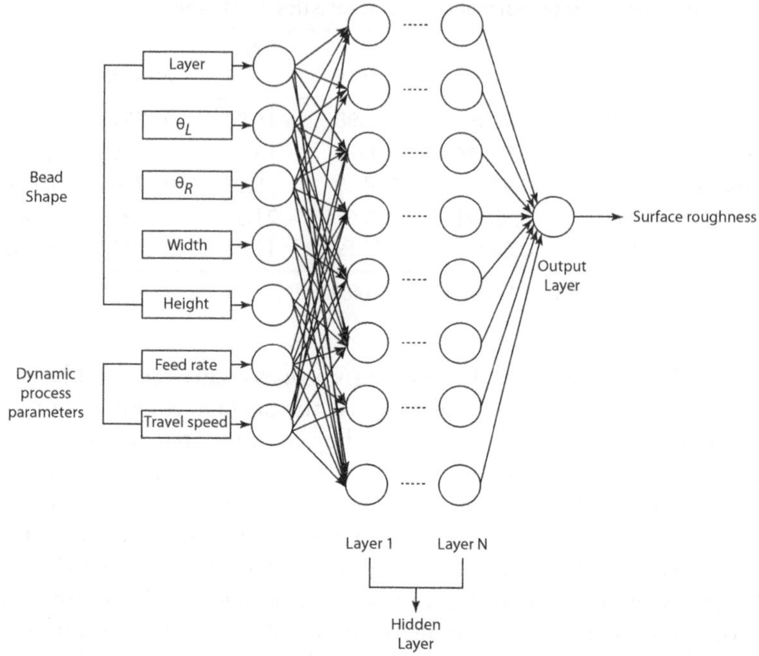

Figure 3.7 Schematic diagram of the deep neural network for surface roughness prediction.

intrinsic to robotic WADED and the resultant layer roughness. The findings emphasize that both methodologies yield favorable results in this context. Noteworthy, however, is the increased performance exhibited by the random forest approach compared to the MLP counterpart. The developed random forest model attains a notable precision in forecasting the roughness of beads based on the provided input parameters. The average error rate for validation and testing phases remains below 6%, signifying the model's robustness. A salient advantage of the random forest methodology lies in its computational efficiency, which paves the way for its potential extension in online roughness modeling. This facet, in turn, facilitates the integration of the model into quality control procedures, enhancing the overall quality assurance framework.

3.7 SUMMARY AND SCOPE OF RESEARCH

3.7.1 Summary

The field of WADED has been a focal point of intensive research, delving into the intricate relationships between a diverse array of process parameters and the resultant surface roughness characteristics of the fabricated structures. Multiple researchers have contributed to shedding light on the multifaceted dynamics governing surface roughness. Parameters such as heat input, wire feed speed, travel speed, interlayer temperature, and current are pivotal determinants that significantly influence the surface finish quality. The intricate interactions among these factors highlight the complexity of optimizing surface characteristics. Innovative cooling strategies have been devised to address surface quality and wall-widening challenges. Various cooling techniques have been introduced and explored, including integrated heat exchangers, forced dry air, and near-immersion active cooling. These strategies effectively mitigate heat accumulation, substantially improving surface roughness and overall part quality. The amalgamation of these research endeavors provides a comprehensive panorama of the influential factors impacting surface roughness in the context of WADED. As the field advances, a holistic understanding of these parameters will steer the development of optimized deposition strategies, thereby contributing to enhanced surface finish and broader applications within additive manufacturing. Within surface quality and texture analysis, the separation distance between adjacent beads has emerged as a critical factor that profoundly influences the resultant surface attributes. The interplay between separation distance, bead width, and height orchestrates the topographical features of the fabricated surface. Detailed analysis reveals a pivotal juncture where a carefully calibrated reduction in separation distance leads to an augmented overlap region, ultimately merging with the valley region to produce an optimal, level surface. However, pursuing minimal separation distances may result in an undesirable outcome characterized by excessive overlap, disrupting layer thickness and surface uniformity. This exploration introduces the concept of the ideal separation distance, exemplified by the FOM and TOM, as a comprehensive framework for comprehending and predicting surface characteristics. Diverse machine learning techniques have also been harnessed to enhance vertical and lateral surface roughness in components produced through WADED. The implementation of the ANFIS, SVR, and ELM methodologies has enabled predictive modeling of surface roughness. The ANFIS model, which amalgamates fuzzy logic and neural networks, has exhibited remarkable predictive accuracy, with the potential for further optimization using metaheuristic algorithms such as PSO and GA. Gaussian process regression has been employed to model and forecast surface roughness, yielding impressive predictive deviations. Additionally, an MLP approach has demonstrated the capacity for accurate prediction even with reduced datasets. Furthermore, a comparative study involving random forest and MLP techniques for surface roughness prediction showcased favorable outcomes, with the random forest model outperforming the MLP model. The computational efficiency of the random forest technique enhances its suitability for online roughness modeling, thereby contributing to heightened quality control measures.

3.7.2 Scope of research

The current body of research has made significant strides in advancing our comprehension of surface roughness in the context of WADED. However, there remain several notable gaps that necessitate further investigation. First, there is a need to delve deeper into the interactions among multiple parameters, extending beyond individual or pairwise effects, to uncover complex synergies that could offer novel insights into optimizing surface roughness. Additionally, the impact of material properties and microstructural changes induced by the deposition process on surface roughness requires more exploration. A thorough understanding of these underlying mechanisms can enhance predictive models for surface quality.

Furthermore, there is a call to pay greater attention to incorporating advanced computational techniques, such as machine learning and artificial intelligence, for predicting and optimizing surface roughness in WADED. These modern methodologies have the potential to reveal intricate patterns and relationships that may not be discernible through traditional analyses. The translation of laboratory findings into real-world industrial applications necessitates robust validation methods. Establishing standardized evaluation approaches for surface roughness, particularly in complex geometries and large-scale components, is crucial to ensure the practical effectiveness of proposed optimization strategies.

Despite progress, gaps persist in the current knowledge, including applying the TOM in scenarios involving intricate bead interactions and varying process conditions. Extending TOM to encompass diverse bead profiles beyond the parabolic cross-section and integrating it with evolving welding techniques and materials presents a promising avenue for further research. Validating the recalibrated critical offset distance through empirical studies across practical welding scenarios will enhance its theoretical basis and practical applicability.

In the realm of machine learning techniques, while significant strides have been taken to predict surface roughness in WADED, critical gaps remain:

i. Comprehensive multiobjective optimization: Current efforts often focus on isolated parameter effects, leaving a gap in holistic multiobjective optimization that considers a broader array of intrinsic and extrinsic variables in the WADED process.
ii. Algorithm fusion and hybrid approaches: Combining machine learning algorithms or hybrid methodologies could enhance predictive accuracy and bridge existing gaps.
iii. Real-time monitoring and control: Although progress has been made in online modeling, there is a need for real-time monitoring and control of the WADED process, adapting to changing conditions during fabrication.

iv. Model generalization and transferability: Extending research beyond specific material systems or process parameters to explore the generalization and transferability of machine learning models is crucial.
v. Uncertainty quantification: Incorporating uncertainty quantification techniques into predictive models could provide a more comprehensive understanding of the reliability of surface roughness predictions.

In most cases, surface roughness has been assessed using vision-based techniques like optical microscopes and laser-based profilometers. The collected data is then processed into point clouds, forming the basis for evaluating surface roughness. Alternative methods, such as acoustic sensor-based assessments and the inductance method, have also been explored. Although efforts have been made to enhance surface roughness prediction and optimization, research often focuses on a limited subset of parameters. Yet, various intrinsic and extrinsic variables significantly influence surface roughness in WADED components. Consequently, there is a demand for comprehensive multiobjective optimization strategies that encompass input parameters and manufacturing tactics. Historically, a limited range of machine learning (ML) and artificial intelligence (AI) techniques have been employed for predicting WADED surface roughness (Figure 3.6). However, recent times have witnessed an array of ML and AI techniques tailored for manufacturing processes (Figure 3.8). This highlights the urgent need to establish a comprehensive framework incorporating diverse ML and AI methodologies. This integrated approach would enable manufacturers

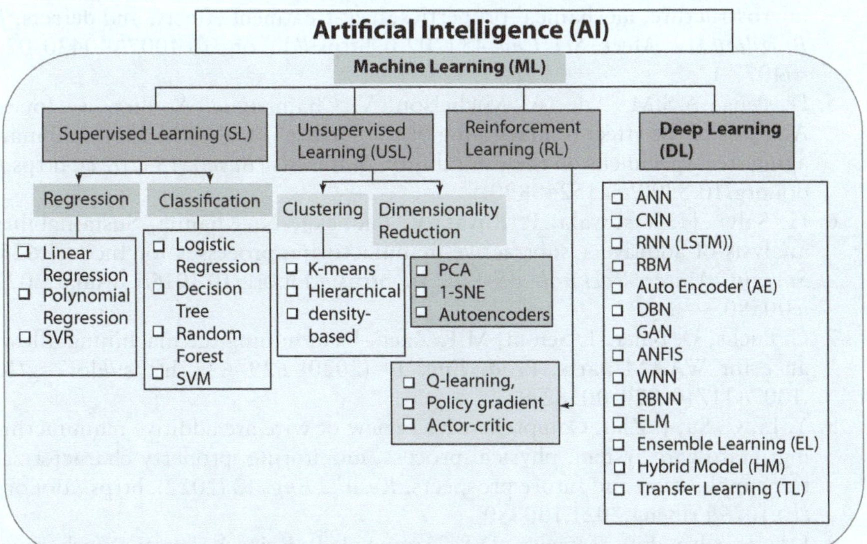

Figure 3.8 Machine learning and artificial intelligence techniques used in different aspects of manufacturing process optimization and prediction.

to confidently predict surface roughness based on deposition system characteristics.

ACKNOWLEDGMENT

The authors acknowledge CSIR-Central Mechanical Engineering Research Institute (CMERI), Durgapur, for infrastructural support.

REFERENCES

1. B. Tomar, S. Shiva, T. Nath, A review on wire arc additive manufacturing: Processing parameters, defects, quality improvement and recent advances, *Mater. Today Commun.* 31 (2022) 103739. https://doi.org/10.1016/j.mtcomm.2022.103739.
2. M.R. Zahidin, F. Yusof, S.H. Abdul Rashid, S. Mansor, S. Raja, M.F. Jamaludin, Y.H. Manurung, M.S. Adenan, N.I. Syahriah Hussein, Research challenges, quality control and monitoring strategy for Wire Arc Additive Manufacturing, *J. Mater. Res. Technol.* 24 (2023) 2769–2794. https://doi.org/10.1016/j.jmrt.2023.03.200.
3. W.S. James, S. Ganguly, G. Pardal, Selection and performance of AM superalloys for high-speed flight environments, *Int. J. Adv. Manuf. Technol.* 122 (2022) 2319–2327. https://doi.org/10.1007/s00170-022-10005-9.
4. M. Bhuvanesh Kumar, P. Sathiya, S.M. Senthil, A critical review of wire arc additive manufacturing of nickel-based alloys: Principles, process parameters, microstructure, mechanical properties, heat treatment effects, and defects, *J. Brazilian Soc. Mech. Sci. Eng.* 45 (2023). https://doi.org/10.1007/s40430-023-04077-1.
5. D. Peng, A.S.M. Ang, A. Michelson, V. Champagne, A. Birt, R. Jones, Analysis of the effect of machining of the surfaces of WAAM 18Ni 250 maraging steel specimens on their durability, *Materials (Basel).* 15 (2022). https://doi.org/10.3390/ma15248890.
6. H. Salvi, H. Vesuwala, P. Raval, V. Badheka, N. Khanna, Sustainability analysis of additive + subtractive manufacturing processes for Inconel 625, *Sustain. Mater. Technol.* 35 (2023). https://doi.org/10.1016/j.susmat.2023.e00580.
7. C. Fuchs, D. Baier, T. Semm, M.F. Zaeh, Determining the machining allowance for WAAM parts, *Prod. Eng.* 14 (2020) 629–637. https://doi.org/10.1007/s11740-020-00982-9.
8. Y. Li, C. Su, J. Zhu, Comprehensive review of wire arc additive manufacturing: Hardware system, physical process, monitoring, property characterization, application and future prospects, *Results Eng.* 13 (2022). https://doi.org/10.1016/j.rineng.2021.100330.
9. L.J. da Silva, F.R. Teixeira, D.B. Araújo, R.P. Reis, A. Scotti, Work envelope expansion and parametric optimization in WAAM with relative density and surface aspect as quality constraints: The case of Al5Mg thin walls with active cooling, *J. Manuf. Mater. Process.* 5 (2021). https://doi.org/10.3390/jmmp5020040.

10. K. Samadian, W. de Waele, Fatigue crack growth model incorporating surface waviness for wire+arc additively manufactured components, *Procedia Struct. Integr.* 28 (2020) 1846–1855. https://doi.org/10.1016/j.prostr.2020.11.008.
11. D. Ding, Z. Pan, D. Cuiuri, H. Li, Wire-feed additive manufacturing of metal components: Technologies, developments and future interests, *Int. J. Adv. Manuf. Technol.* 81 (2015) 465–481. https://doi.org/10.1007/s00170-015-7077-3.
12. Y. Yehorov, L.J. da Silva, A. Scotti, Balancing WAAM production costs and wall surface quality through parameter selection: A case study of an Al-Mg5 alloy multilayer-non-oscillated single pass wall, *J. Manuf. Mater. Process.* 3 (2019). https://doi.org/10.3390/jmmp3020032.
13. J. Xiong, Y.J. Li, Z.Q. Yin, H. Chen, Determination of surface roughness in wire and arc additive manufacturing based on laser vision sensing, *Chinese J. Mech. Eng. (English Ed.)* 31 (2018) 1–7. https://doi.org/10.1186/s10033-018-0276-8.
14. J. Xiong, Y. Li, R. Li, Z. Yin, Influences of process parameters on surface roughness of multi-layer single-pass thin-walled parts in GMAW-based additive manufacturing, *J. Mater. Process. Technol.* 252 (2018) 128–136. https://doi.org/10.1016/j.jmatprotec.2017.09.020.
15. K. Biester, M. Lammers, J. Hermsdorf, S. Kaierle, Influence of the laser position in laser-assisted WAAM process on weld bead shape and surface properties, *Procedia CIRP.* 111 (2022) 190–195. https://doi.org/10.1016/j.procir.2022.08.044.
16. C. Xia, Z. Pan, J. Polden, H. Li, Y. Xu, S. Chen, Modelling and prediction of surface roughness in wire arc additive manufacturing using machine learning, *J. Intell. Manuf.* 33 (2022) 1467–1482. https://doi.org/10.1007/s10845-020-01725-4.
17. H. Chen, A. Yaseer, Y. Zhang, Top surface roughness modeling for robotic wire arc additive manufacturing, *J. Manuf. Mater. Process.* 6 (2022). https://doi.org/10.3390/jmmp6020039.
18. C. Zhang, Z. Qiu, H. Zhu, Z. Wang, O. Muránsky, M. Ionescu, Z. Pan, J. Xi, H. Li, On the effect of heat input and interpass temperature on the performance of inconel 625 alloy deposited using wire arc additive manufacturing–cold metal transfer process, *Metals (Basel).* 12 (2022). https://doi.org/10.3390/met12010046.
19. F. Jiang, L. Sun, R. Huang, H. Jiang, G. Bai, X. Qi, C. Liu, Y. Su, C. Guo, J. Wang, Effects of heat input on morphology of thin-wall components fabricated by wire and arc additive manufacturing, *Adv. Eng. Mater.* 23 (2021) 1–12. https://doi.org/10.1002/adem.202001443.
20. V.T. Le, M.C. Bui, T.D. Nguyen, V.A. Nguyen, V.C. Nguyen, On the connection of the heat input to the forming quality in wire-and-arc additive manufacturing of stainless steels, *Vacuum.* 209 (2023) 111807. https://doi.org/10.1016/j.vacuum.2023.111807.
21. M. Dinovitzer, X. Chen, J. Laliberte, X. Huang, H. Frei, Effect of wire and arc additive manufacturing (WAAM) process parameters on bead geometry and microstructure, *Addit. Manuf.* 26 (2019) 138–146. https://doi.org/10.1016/j.addma.2018.12.013.

22. F.R. Teixeira, F.M. Scotti, V.L. Jorge, A. Scotti, Combined effect of the interlayer temperature with travel speed on features of thin wall WAAM under two cooling approaches, *Int. J. Adv. Manuf. Technol.* (2023) 273–289. https://doi.org/10.1007/s00170-023-11105-w.
23. Y.B. Xiong, D.X. Wen, Z.Z. Zheng, J.J. Li, Effect of interlayer temperature on microstructure evolution and mechanical performance of wire arc additive manufactured 300M steel, *Mater. Sci. Eng. A.* 831 (2022) 142351. https://doi.org/10.1016/j.msea.2021.142351.
24. A. Dash, L. Squires, J.D. Avila, S. Bose, A. Bandyopadhyay, Influence of active cooling on microstructure and mechanical properties of wire arc additively manufactured mild steel, *Front. Mech. Eng.* 9 (2023) 1–11. https://doi.org/10.3389/fmech.2023.1130407.
25. V.T. Le, D.S. Mai, H. Paris, Influences of the compressed dry air-based active cooling on external and internal qualities of wire-arc additive manufactured thin-walled SS308L components, *J. Manuf. Process.* 62 (2021) 18–27. https://doi.org/10.1016/j.jmapro.2020.11.046.
26. V.T. Le, D.S. Mai, Q.H. Hoang, Effects of cooling conditions on the shape, microstructures, and material properties of SS308L thin walls built by wire arc additive manufacturing, *Mater. Lett.* 280 (2020) 128580. https://doi.org/10.1016/j.matlet.2020.128580.
27. F.M. Scotti, F.R. Teixeira, L.J. da Silva, D.B. de Araújo, R.P. Reis, A. Scotti, Thermal management in WAAM through the CMT Advanced process and an active cooling technique, *J. Manuf. Process.* 57 (2020) 23–35. https://doi.org/10.1016/j.jmapro.2020.06.007.
28. J. Long, M. Wang, W. Zhao, X. Zhang, Y. Wei, W. Ou, High-power wire arc additive manufacturing of stainless steel with active heat management, *Sci. Technol. Weld. Join.* 27 (2022) 256–264. https://doi.org/10.1080/13621718.2022.2045127.
29. S. Suryakumar, K.P. Karunakaran, A. Bernard, U. Chandrasekhar, N. Raghavender, D. Sharma, Weld bead modeling and process optimization in Hybrid Layered Manufacturing, *CAD Comput. Aided Des.* 43 (2011) 331–344. https://doi.org/10.1016/j.cad.2011.01.006.
30. J. Xiong, G. Zhang, J. Hu, L. Wu, Bead geometry prediction for robotic GMAW-based rapid manufacturing through a neural network and a second-order regression analysis, *J. Intell. Manuf.* 25 (2014) 157–163. https://doi.org/10.1007/s10845-012-0682-1.
31. Y. Cao, S. Zhu, X. Liang, W. Wang, Overlapping model of beads and curve fitting of bead section for rapid manufacturing by robotic MAG welding process, *Robot. Comput. Integr. Manuf.* 27 (2011) 641–645. https://doi.org/10.1016/j.rcim.2010.11.002.
32. J. Xiong, G. Zhang, H. Gao, L. Wu, Modeling of bead section profile and overlapping beads with experimental validation for robotic GMAW-based rapid manufacturing, *Robot. Comput. Integr. Manuf.* 29 (2013) 417–423. https://doi.org/10.1016/j.rcim.2012.09.011.
33. D. Ding, Z. Pan, D. Cuiuri, H. Li, A multi-bead overlapping model for robotic wire and arc additive manufacturing (WAAM), *Robot. Comput. Integr. Manuf.* 31 (2015) 101–110. https://doi.org/10.1016/j.rcim.2014.08.008.

34. Y. Tiwari, M. Mukherjee, D. Chatterjee, R. Manivannan, Influence of interlayer rotation in parallel deposition strategies on the microstructure, texture, and mechanical behaviour of Inconel-625 during directed energy deposition, *Mater. Charact.* 197 (2023) 112711. https://doi.org/10.1016/j.matchar.2023.112711.
35. S.A. Banaee, A. Kapil, F. Marefat, A. Sharma, Generalised overlapping model for multi-material wire arc additive manufacturing (WAAM), *Virtual Phys. Prototyp.* 18 (2023). https://doi.org/10.1080/17452759.2023.2210541.
36. Z. Zhang, J. Yan, X. Lu, T. Zhang, H. Wang, Optimization of porosity and surface roughness of CMT-P wire arc additive manufacturing of AA2024 using response surface methodology and NSGA-II, *J. Mater. Res. Technol.* 24 (2023) 6923–6941. https://doi.org/10.1016/j.jmrt.2023.04.259.
37. A. Yaseer, H. Chen, B. Zhang, Predicting layer roughness with weaving path in robotic wire arc additive manufacturing using multilayer perceptron, *2021 IEEE 11th Annu. Int. Conf. CYBER Technol. Autom. Control. Intell. Syst. CYBER* 2021. (2021) 61–66. https://doi.org/10.1109/CYBER53097.2021.9588272.
38. A. Yaseer, H. Chen, Machine learning based layer roughness modeling in robotic additive manufacturing, *J. Manuf. Process.* 70 (2021) 543–552. https://doi.org/10.1016/j.jmapro.2021.08.056.

Chapter 4

Fabrication of super-polished aluminium mandrels for x-ray optics

J.V. Sai Prasanna Kumar and Dharmalingam G

4.1 INTRODUCTION

Telescopes are tools to witness a celestial event, planetary transits, search for aliens on Mars or hunt for exoplanets. Telescopes are constructed differently to suit different applications. Their major components are mirrors for focusing, the detecting stage, transmission stage and the images relayed to ground stations and their subsequent analysis. Working on this premise to detect cosmic rays, the detecting mirrors require larger surface areas to gather light and detectors to relay the images spied upon. This indicates that most of the telescopes used in the visible wavelength are particularly not suitable in other wavelength regimes. It is a fact that the reflectivity at the surface is almost negligible, but conversely the total external reflection occurs and x-rays can be reflected back, according to Snell's law, at a critical angle, termed as the grazing angle. This reflection generally occurs at low energies; this phenomenon is called grazing. This concept is exploited by x-ray mirrors and to extend further, mirror shell surfaces can be a set of paraboloids or hyperboloids arranged coaxially or confocally to increase the surface areas without being too heavy. This type of arrangement is known as Wolter-I mirrors.

X-ray optics plays a vital role in high-energy detection in a wide range of applications. For instance, the X-ray Timing and Polarimetry mission space observatory [1, 2] led by the Chinese government, is to be launched in 2027 and is the future of x-ray satellites. Since the 1970s, several research institutions have conducted profound research on fabricating the Wolter-I type focusing mirror. The steps involved in manufacturing the mirrors are direct polishing [3], splicing [4, 5], thermal collapse [6] and the replication technique [7–12]. Presently the manufacturing process relies on precision electroforming techniques, introduced in the 1990s [13], and since then several satellites launched by ESA and NASA have relied on this method [14–19]. Electroforming happens to be the most reliable and sought-after method of developing x-ray mirrors [20].

The electroforming method has several advantages, namely, it is precise, cost effective and reliable. However, the processing of demoulding

 DOI: 10.1201/9781003713364-4

the mirror from the mandrel is a tedious process [21]. Several methods, like passivation [23] or adding a TiN coating [23, 24] can reduce the binding force. Sometimes, a diamond-like carbon (DLC) coating is also used to reduce the demoulding force.

India's AstroSat observatory observes cosmic objects in a band of x-rays in the energy range from 0.3 to 100 keV. Launched by the Indian Space Research Organisation (ISRO) in 2015, this observatory is positioned at 650 km above the Earth's surface and has an orbital period of 98 minutes. The instruments that are carried aboard are a particle monitor, telluride imager, scanning sky monitor and xenon proportional counters. The next stage is the reflectors of mirrors for this mission; here the mirrors are assembled in multitude and in a segmented way, mostly shells of conic sections. Glass and aluminium are used as reflecting shell elements. The shells are nested together in the mirror's module, building on weight saving and low cost. The telescope is shown in Figure 4.1.

The international status in developing such telescopes is very much like our endeavour to unravel the mysteries of deep space and enrich our knowledge of cosmic dynamics. The mirror module is represented in Figure 4.2.

India's Polar Satellite Launch Vehicle is shown in Figure 4.3, and in its 31st flight (PSLV-C30) launched 1515 kg AstroSat into a 650 km orbit of 6 degrees inclination to the equator. PSLV-C30 was launched from the First Launch Pad (FLP) of Satish Dhawan Space Centre (SDSC) SHAR, Sriharikota, on 28 September 2015.

The objectives of the AstroSat mission are to design, develop, realise and launch a multi-wavelength astronomy satellite for studying the cosmic sources simultaneously over a wide range of the electromagnetic spectrum, that is, from optical, ultraviolet (UV) to high-energy x-rays. The AstroSat unit is shown in Figure 4.4.

The AstroSat has several segments, namely, mirror segment, payloads like the cameras, ground relay segments, a CFRP tube which houses the baffle assembly and protects the optics assembly from the sun, and several other sophisticated instruments. The most important segment in any telescope is the mirrors. Traditionally, glass is one of the best mirror materials and is still being used, but other materials are aluminium, beryllium and carbon fibre. At the same time, the mirrors are given reflective coating so that their optical performance is enhanced. Gold happens to be the most sought-after coating material because of its inertness. The method and manufacture of mirrors have remained the same for the past several decades. Right now, we have access to sophisticated machines like single-point diamond turning machines, which are used to maintain submicron surface accuracies, and in situ metrology equipment to check the surface roughness, like non-contact profilometers, atomic force and confocal microscopes, enabling us to maintain the surface irregularities at bay. In the case of the telescopes which are used to study cosmic rays and relay the information/images of a nebula and distant stars, soft x-ray telescopes can be employed. All astrophysicists

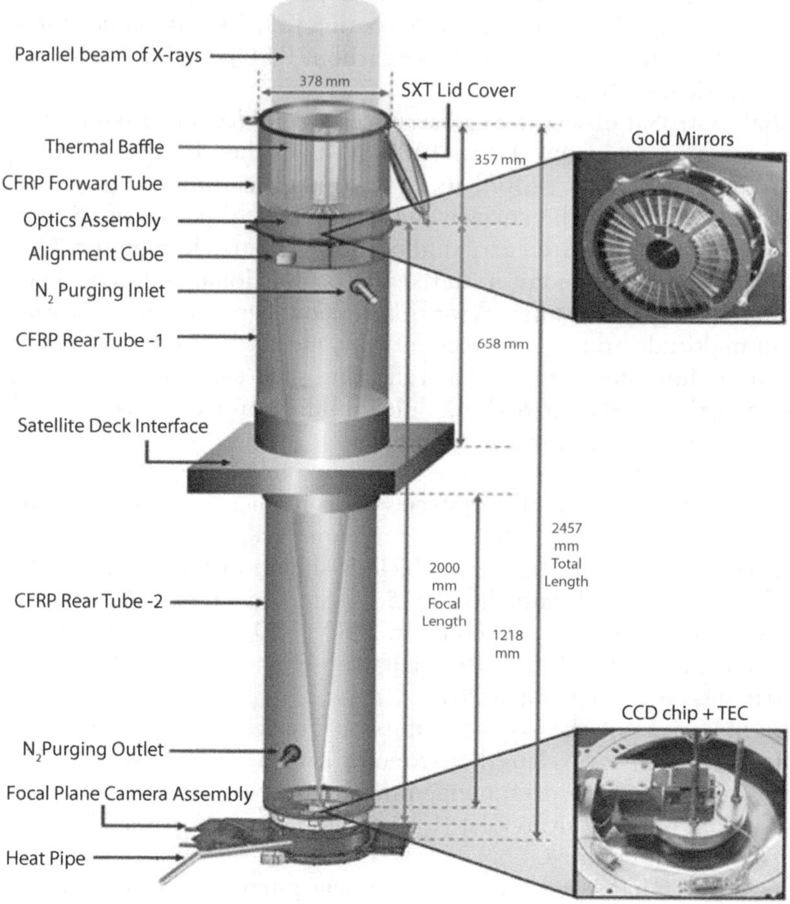

Figure 4.1 Schematic of the telescope developed by ISRO. (From [22], Navalkar et al., *J. Astrophys. Astron.*, 2021, 42(2).)

strive to develop a mirror that is of low cost, extremely powerful, excellent performance and lightweight. But for soft x-ray telescopes, more surface area is required without sacrificing the other parameters.

4.2 MATERIALS AND METHODS

Ideally, aerospace-grade aluminium can lend itself to surface manipulation and can be super polished to the requisite nanometre level surface finish measured in RMS. Aluminium is a relatively soft, easy-to-machine metal, so it lends itself to an excellent polished surface and can be manipulated into

Figure 4.2 Mirror module. (From [22], Navalkar et al., *J. Astrophys. Astron.*, 2021, 42(2).)

Figure 4.3 Polar Satellite Launch Vehicle. (From ISRO website.)

Figure 4.4 AstroSat integration unit. (From ISRO website.)

a mirror-like appearance. A 400 mm x 600 mm x 30 mm aluminium 6061, an aerospace-grade plate, has been procured (Figure 4.5) and machined to the requisite shell element segment (Figure 4.6).

Al-6061 aluminium alloy is a general-purpose structural alloy developed by Alcoa in 1935 with temper designations such as 6061-T4, T451, Al 6061-T6 and 6061-T651. It is one of the most widely used alloys. The main alloying elements are magnesium (Mg) and silicon (Si). Aluminium 6061 has good corrosion resistance, mechanical properties, formability, weldability and machinability. Generally suitable for medium to high strength requirements, it has good toughness. To machine the sample on hand, the following was adopted on the 5-axis milling VMC with the following specifications: X-axis movement, 800 mm; Y-axis movement, 900 mm; spindle speed, up to 12000 rev/min; feed rate, 0.001–20 m/min; nose tool radius, up to 0.01 mm. Tool changers and automatic controls are also available. The evaluation of the surface roughness was performed off-line on the sample using a contact-based stylus profilometer, whose schematic is shown in Figure 4.7, from the Mitutoyo company.

Figure 4.5 Flat 6061-T aluminium plate.

Figure 4.6 Aluminium machined as shell element.

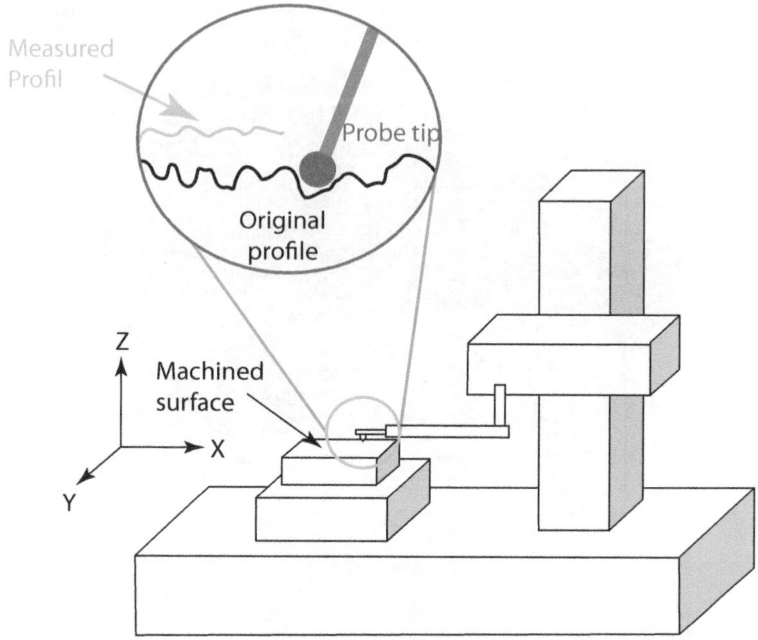

Figure 4.7 Stylus-based contact profilometer.

This instrument can evaluate the surface accuracies over a length of 4 mm; the stylus has a tip radius of 5 µm, while the probe is made from diamond. It can be used to measure the surface roughness up to 0.001 µm (1 nm). All the measurements were made at 20°C. The sample was initially milled using the VMC, which was set up at a feed rate of 100 mm/min; the spindle speed was 500 rpm, while the nose radius was set to 1.5 mm. The sample from the aluminium block was chipped off to a quadrant of a circle with an arc radius of 37.5 mm, making 90 degrees the angle of the arc at the centre, while the length of the arc was about 250 mm (Figure 4.8). The surface roughness was measured by a handheld device at one particular location and is reported as 31 µm.

The next step was to set the milling machine at a feed rate of 800 mm/min and spindle speed of 1000 rev/min, with a depth of cut maintained at 2 mm. The sample assumed a matte-like appearance, with numerous tool lines running parallel to the surface appeared (Figure 4.9). The Ra was measured over a length of 4 mm and it was found to be 0.631 µm (631 nm) (Figure 4.10).

The feed rate, depth of the cut and spindle speeds were set at the following specifications and the milling process continued, all the while using standard coolants lest the surface deforms under thermal loads: feed rate, 1500 mm/min; spindle speed, 6000 rpm; depth of the cut, 0.5 mm; nose radius of the tool, 0.5 mm (Figure 4.11). Once again the surface roughness

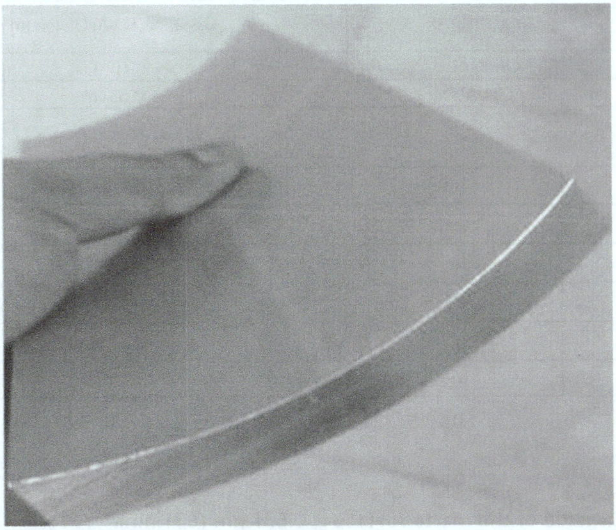

Figure 4.8 Initial milling and generation of the shape for the mandrel.

Figure 4.9 Numerous tool lines present generated a matte finish on the mandrel.

was evaluated at the ambient conditions over a length of 4 mm; it was found to be 0.209 µm (209 nm) (referenced in Figure 4.12).

Here also numerous tool lines appeared and were running parallel to the surface of the sample. Also, at some point during machining, a slight warping appeared, but that warpage was removed by applying a fine grit emery

PARAMETER	OBSERVED VALUE IN μm
Ra NOMINAL	2μm
Ra ACTUAL	0.631m

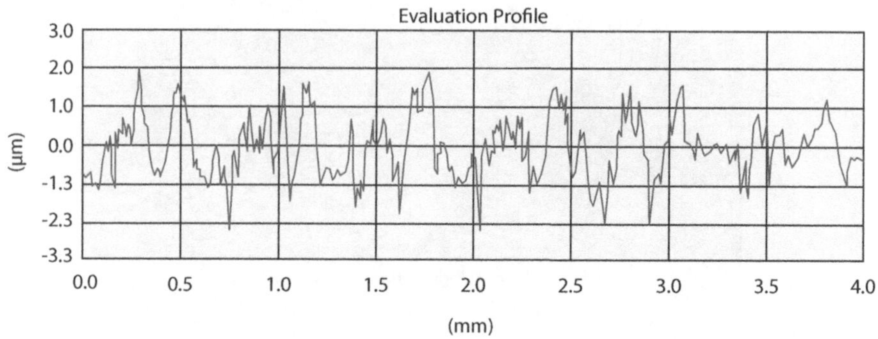

Figure 4.10 Ra value of the matte surface is 0.631 μm (631 nm).

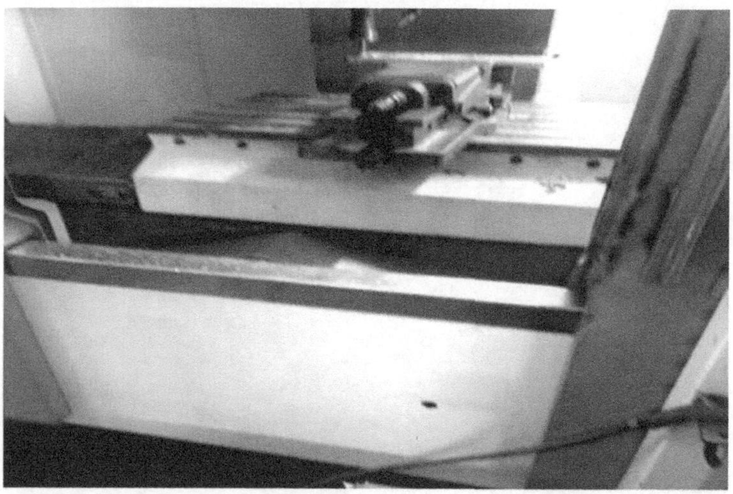

Figure 4.11 Milling operation under process.

paper. The machining operation was continued with the process parameters specified as follows: feed rate, 2000 mm/min; spindle speed, 6500 rpm; depth of cut, 0.25 mm; nose radius of the tool, 0.25 mm. Here, an attempt was made to remove the tool lines, and a glossy finish was obtained. The surface roughness referenced in the figure is 0.068 μm (68 nm), and thus a glossy appearance was obtained (Figures 4.13 and 4.14).

PARAMETER	OBSERVED VALUE IN µm
Ra NOMINAL	2µm
Ra ACTUAL	0.209µm

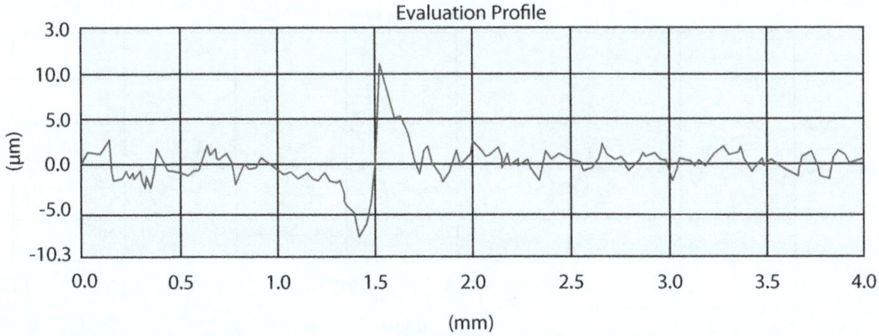

Figure 4.12 Surface roughness evaluated at Ra = 0.209 µm (209 nm).

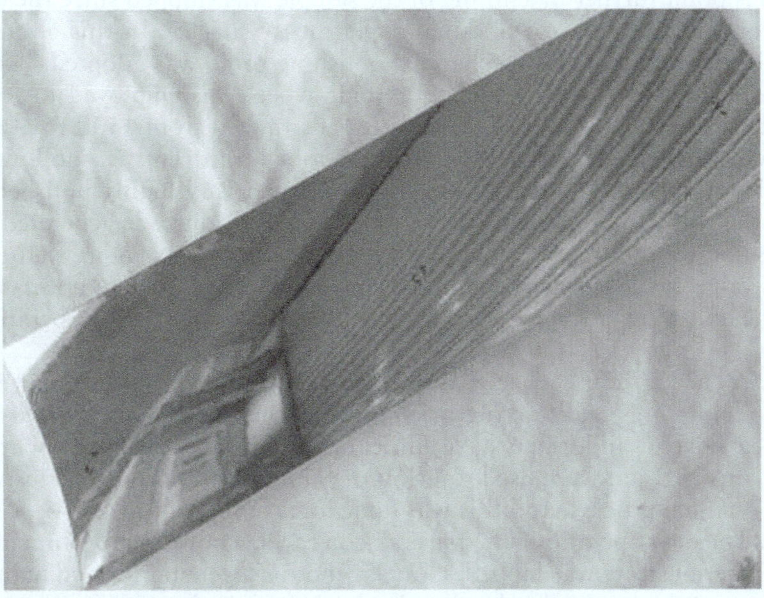

Figure 4.13 Removal of tool lines.

PARAMETER	OBSERVED VALUE IN μm
Ra NOMINAL	μm
Ra ACTUAL	0.68μm

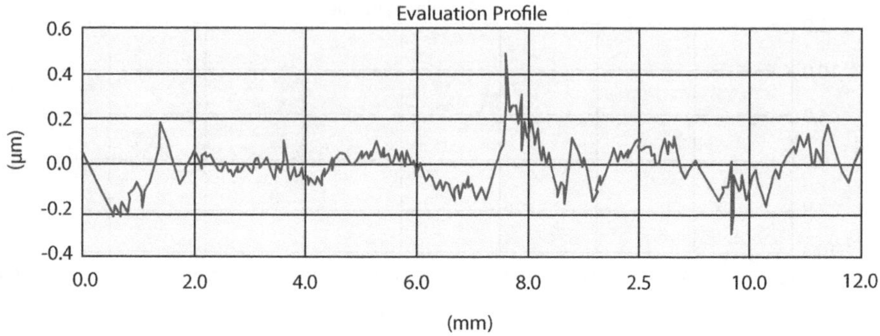

Figure 4.14 Improvement in Ra = 0.068 μm (68 nm) for unpolished surface.

4.3 POLISHING OF THE SAMPLE

At this point further machining of the sample was stopped since the objective of this work was to obtain a super glossy surface. In this attempt, a combined chemical and mechanical polishing route was adopted: initially, a rough polishing with a silicon carbide wheel was employed to remove the burr and micro fissures. Then grit pads ranged from 1500 to 600 grit wheels. The sequence of operations is described next. This was followed by employing diamond-based slurries to generate a very highly polished surface (Figure 4.15) and the final corresponding Ra value was generated from the same procedure described elsewhere (referenced in Figure 4.16). The final value of the surface roughness was found to be 0.0031 μm (3.1 nm).

The polishing of the mandrel had the following sequence of operations:

1. A hard grinding wheel was introduced to grind the surface to even out the fissures and micro cracks. Here in this stage, water was used as a solvent, and it acts as a lubricant. Aluminium oxide nanoparticles of size 2 μm were used to grind the aluminium substrate. This was a manual process, and this was done for 1 hour (Figure 4.15).
2. The next operation was to use silicon carbide powder of 1.5 μm in size; the substrate was lapped. Circular motions were performed manually at very low speeds; the shearing action of the loose aluminium microparticles and the solvent creates a slurry, ensuring a great material removal. Thus, a uniformly smooth surface was obtained. Very light

Figure 4.15 Grinding wheel to even out the micro burrs.

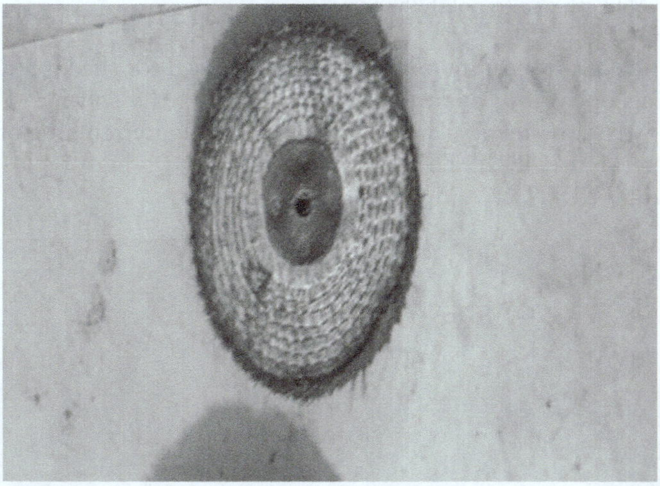

Figure 4.16 Mild grit tool to even out the fissures and surface unevenness, lapping process.

scratches could be visible at this stage. This operation is performed for another hour.

3. The lapping process is continued with aluminium oxide of size 1.5 μm; again, water is used as the lubricant. With the application of the slurry, the wheel having fine grit eases out the scratches; a polished surface starts to show after one hour of buffing (Figure 4.17).

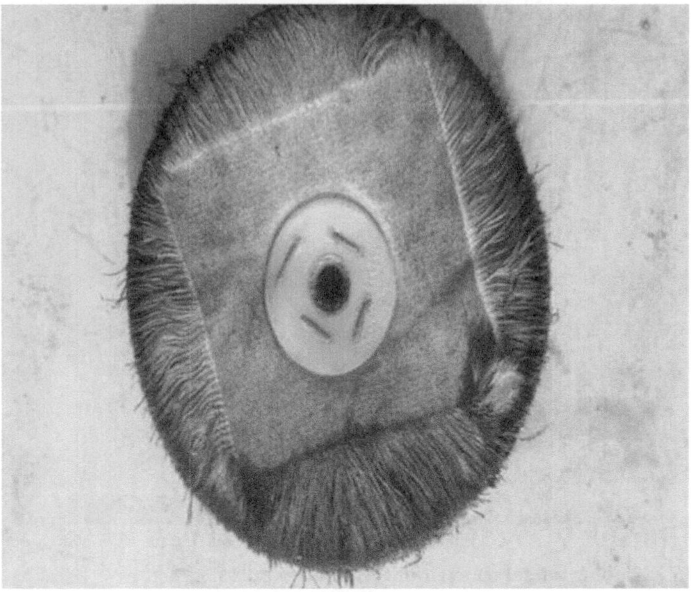

Figure 4.17 Mild grit pad to polish the sample.

4. Optical grade cerium oxide powders are used for the lapping and polishing operation; again, water was used as the solvent. The polishing was done using a fine grit pad by making circular motions for a hour (Figure 4.18). This operation generated a highly glossy surface (Figure 4.19).

Figure 4.18 Diamond slurry to generate a super finish.

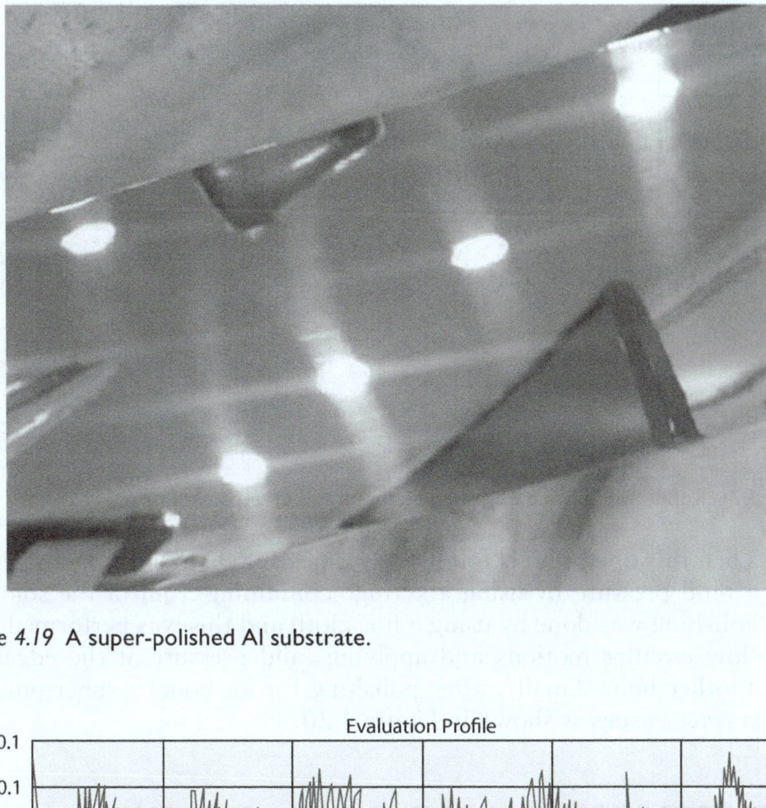

Figure 4.19 A super-polished Al substrate.

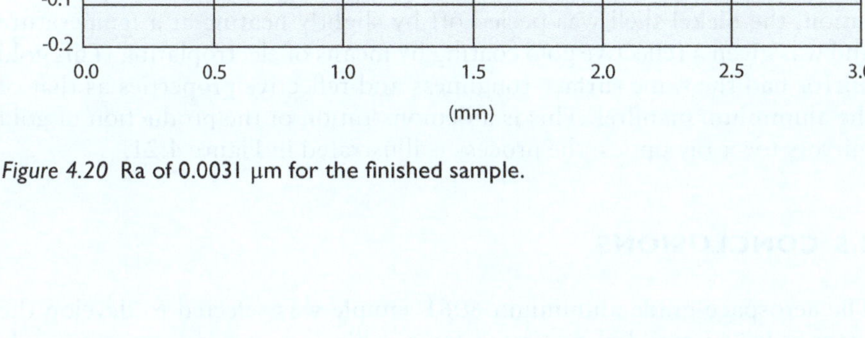

Figure 4.20 Ra of 0.0031 µm for the finished sample.

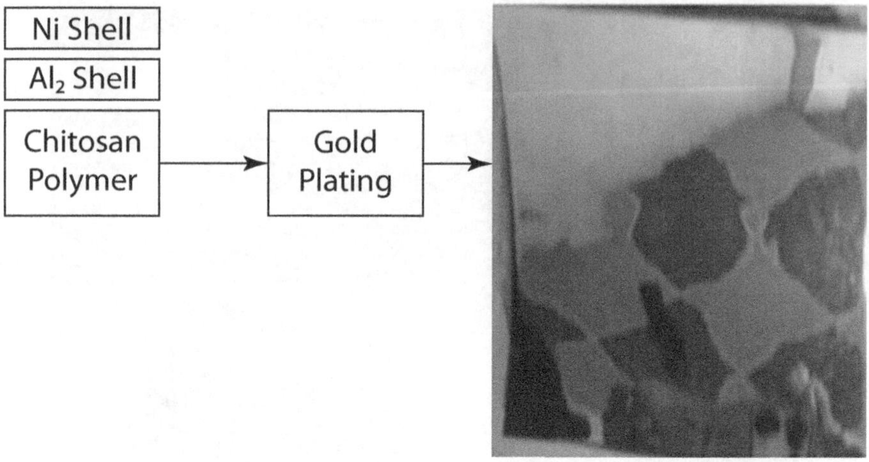

Figure 4.21 Reflective gold-coated shell element and electroforming process.

5. Then this operation is followed by injecting diamond-based slurry at a mild pressure by using a syringe containing 5 ml of the solution. Polishing was done by using a fine cloth and this was performed with slow circular motions and applying mild pressure at the edges for another hour. Finally, after polishing for an hour, a super-polished surface ensues as shown in Figure 4.20.

4.4 ELECTROFORMING OF THE SHELL ELEMENT

Chitosan, an organic polymer, was used to produce a nickel shell element by embossing the aluminium shell on the organic polymer. After polymerisation, the nickel shell was peeled off by slightly heating at a temperature and was given a reflective gold coating by means of electroplating. This gold mirror had the same surface roughness and reflective properties as that of the aluminium mandrel. This is a demonstration of the production of gold mirrors for x-ray optics; the process is illustrated in Figure 4.21.

4.5 CONCLUSIONS

The aerospace-grade aluminium 6061 sample was selected to develop the super-polished mandrel generated from a block to a quadrant of a circle with an angle subtending at the centre of 90 degrees. The radius of the arc was 75 mm, while its length was 254 mm, maintaining a thickness of 10 mm. The sample was initially machined on a VMC which could be operated numerically, and the surface roughness was measured and found to be

31 μm. Further machining and adjusting the process parameters improved the surface roughness to 631 μm, then to 209 μm, generating a matte finish. The machine parameters were refined once again, and after 10 hours of machining, the surface roughness improved to 0.068 μm, which is 68 nm, while generating a glossy surface.

At this stage the machining was stopped because the objective of this work was to generate a super-polished substrate. A mechanical and chemical polishing route was adopted to improve the surface roughness, and the glossy appearance of the sample evolved, rendering it suitable for the next stage of the operation. One final surface roughness evaluated with the procedure illustrated in the earlier section was found to be 0.0031 μm, otherwise 3.1 nm. Further improvement in surface roughness was not attempted. The aluminium mandrel was electroformed into a nickel shell element and was provided with a reflective gold coating. This work was a demonstration of generating a gold mirror for x-ray optics.

ACKNOWLEDGMENT

The authors thank Vel Tech management for providing the financial support (vide slip 0011091/27.10.2022), and the dean of R&D, dean of the School of Mechanical and Construction (SOMC), and the head of the aeronautical engineering department.

CONFLICT OF INTEREST

The author declares that there was not any conflict of interest in pursuing this research work.

REFERENCES

1. Bavdaz, M.; Rando, N.; Wille, E.; Wallace, K.; Shortt, B.; Collon, M.; van Baren, C.; Pareschi, G.; Christensen, F.; Krumrey, M.; et al. ESA-led ATHENA/IXO optics development status. In *Proceedings of the Optics for EUV, X-ray, and Gamma-ray Astronomy V, San Diego, CA, USA, 21–25 August 2011*; SPIE: Bellingham, WA, USA, 2011.
2. Boella, G.; Chiappetti, L.; Conti, G.; Cusumano, G.; Del Sordo, S.; La Rosa, G.; Maccarone, M.C.; Mineo, T.; Molendi, S.; Re, S.; et al. The medium-energy concentrator spectrometer on board the BeppoSAX X-ray astronomy satellite. *Astron. Astrophys. Suppl. Ser.* **1997**, *122*, 327–340.
3. Brinkman, A.C.; Behar, E.; Güdel, M.; Audard, M.; Boggende, A.J.F.D.; Branduardi-Raymont, G.; Cottam, J.; Erd, C.; Herder, J.W.D.; Jansen, F.; et al. First light measurements with the XMM-Newton reflection grating

spectrometers: Evidence for an inverse first ionisation potential effect and anomalous Ne abundance in the Coronae of HR 1099. *Astron. Astrophys.* **2001**, *365*, 75–76.
4. Burrows, D.N.; Hill, J.E.; Nousek, J.A.; Kennea, J.A.; Wells, A.; Osborne, J.P.; Abbey, A.F.; Beardmore, A.; Mukerjee, K.; Short, A.D.T.; et al. The swift X-ray telescope. *Space Sci. Rev.* **2005**, *120*, 165–195.
5. Citterio, O.; Campana, S.; Conconi, P.; Ghigo, M.; Mazzoleni, F.; Poretti, E.; Conti, G.; Cusumano, G.; Sacco, B.; Braeuninger, H.W.; et al. Characteristics of the flight model optics for the JET-X telescope onboard the Spectrum-X-Gamma satellite. In *Proceedings of the Multilayer and Grazing Incidence X-ray/EUV Optics III, Denver, CO, USA, 4–9 August 1996*; SPIE: Bellingham, WA, USA, 1996; Volume 2805, pp. 56–66.
6. De Chambure, D.; Laine, R.; van Katwijk, K.; van Casteren, J.F.; Glaude, P. Status of the X-ray mirror production for the ESA XMM spacecraft. In *Proceedings of the EUV, X-Ray, and Gamma-ray Instrumentation for Astronomy VII, Denver, CO, USA, 4–9 August 1996*; SPIE: Bellingham, WA, USA, 1996; Volume 2808, pp. 362–375.
7. Garmire, G.P.; Bautz, M.W.; Ford, P.G.; Nousek, J.A.; Ricker, G.R., Jr. Advanced CCD imaging spectrometer (ACIS) instrument on the Chandra X-ray observatory. In *Proceedings of the X-ray and Gamma-ray Telescopes and Instruments for Astronomy, Waikoloa, HW, USA, 22–28 August 2002*; SPIE: Bellingham, WA, USA, 2002; Volume 4851, pp. 28–45.
8. Gendreau, K.C.; Arzoumanian, Z.; Adkins, P.W.; Albert, C.L.; Anders, J.F.; Aylward, A.T.; Baker, C.L.; Balsamo, E.R.; Bamford, W.A.; Benegalrao, S.S.; et al. The Neutron star Interior Composition Explorer (NICER): Design and development. In *Proceedings of the Space Telescopes and Instrumentation 2016: Ultraviolet to Gamma ray, Edinburgh, UK, 26 June–1 July 2016*; SPIE: Bellingham, WA, USA, 2016.
9. Grant, C.E.; Bautz, M.W.; Ford, P.G.; Plucinsky, P.P. Fifteen years of the Advanced CCD Imaging Spectrometer. In *Proceedings of the Space Telescopes and Instrumentation 2014: Ultraviolet to Gamma ray, Montréal, QC, Canada, 22–27 June 2014*; SPIE: Bellingham, WA, USA, 2014; Volume 9144.
10. Harrison, F.A.; Boggs, S.; Christensen, F.; Craig, W.; Hailey, C.; Stern, D.; Zhang, W.; Angelini, L.; An, H.; Bhalerao, V.; et al. The Nuclear Spectroscopic Telescope Array (NuSTAR). In *Proceedings of the Space Telescopes and Instrumentation 2010: Ultraviolet to Gamma Ray, 27 June–2 July 2010*; SPIE: Bellingham, WA, USA, 2010.
11. Koglin, J.E.; An, H.; Blaedel, K.L.; Brejnholt, N.F.; Christensen, F.E.; Craig, W.W.; Decker, T.A.; Hailey, C.J.; Hale, L.C.; Harrison, F.A.; et al. NuSTAR hard X-ray optics design and performance. In *Proceedings of the Optics for EUV, X-ray, and Gamma-ray Astronomy IV, San Diego, CA, USA, 2–6 August 2009*; SPIE: Bellingham, WA, USA, 2019.
12. O'Dell, S.L.; Atkins, C.; Broadway, D.M.; Elsner, R.F.; Gaskin, J.A.; Gubarev, M.V.; Kilaru, K.; Kolodziejczak, J.J.; Ramsey, B.D.; Roche, J.M.; et al. X-ray optics at Nasa Marshall Space Flight Center. In *Proceedings of the EUV and X-ray Optics: Synergy between Laboratory and Space IV, Prague, Czech Republic, 13–16 April 2015*; SPIE: Bellingham, WA, USA, 2015; Volume 9510, pp. 7–20.

13. Parmar, A.N.; Martin, D.D.; Bavdaz, M.; Favata, F.; Kuulkers, E.; Vacanti, G.; Lammers, U.; Peacock, A.; Taylor, B.G. The low-energy concentrator spectrometer on-board the BeppoSAX X-ray astronomy satellite. *Astron. Astrophys. Suppl. Ser.* **1997**, *122*, 309–326.
14. Peres, C.B.; Fabian, A.C.; Edge, A.C.; Allen, S.W.; Johnstone, R.M.; White, D.A. A ROSAT study of the cores of clusters of galaxies—I. Cooling flows in an X-ray flux-limited sample. *Mon. Not. R. Astron. Soc.* **1998**, *298*, 416–432.
15. Polletta, M.; Tajer, M.; Maraschi, L.; Trinchieri, G.; Lonsdale, C.J.; Chiappetti, L.; Andreon, S.; Pierre, M.; Le Fevre, O.; Zamorani, G.; et al. Spectral energy distributions of hard X-ray selected active galactic nuclei in the XMM-Newton medium deep survey. *Astrophys. J.* **2007**, *663*, 81.
16. Ravera, L.; Barret, D.; den Herder, J.W.; Piro, L.; Clédassou, R.; Pointecouteau, E.; Peille, P.; Pajot, F.; Arnaud, M.; Pigot, C.; et al. The X-ray integral field unit (X-IFU) for athena. In *Proceedings of the Space Telescopes and Instrumentation 2014: Ultraviolet to Gamma Ray, Montréal, QC, Canada, 22–27 June 2014*; SPIE: Bellingham, WA, USA, 2014.
17. Romaine, S.; Boike, J.; Bruni, R.; Engelhaupt, D.; Gorenstein, P.; Gubarev, M.; Ramsey, B. Mandrel replication for hard X-ray optics using titanium nitride. In *Proceedings of the Optics for EUV, X-ray, and Gamma-ray Astronomy IV, San Diego, CA, USA, 2–6 August 2009*; SPIE: Bellingham, WA, USA, 2009; Volume 7437, pp. 325–332.
18. Romaine, S.; Boike, J.; Bruni, R.; Engelhaupt, D.; Gorenstein, P.; Ramsey, B. Improved release coatings for electroformed X-ray optics. In *Proceedings of the Optics for EUV, X-ray, and Gamma-ray Astronomy V, San Diego, CA, USA, 21–25 August 2011*; SPIE: Bellingham, WA, USA, 2011; Volume 8147, pp. 323–328.
19. Serlemitsos, P.J.; Soong, Y.; Chan, K.W.; Okajima, T.; Lehan, J.P.; Maeda, Y.; Itoh, K.; Mori, H.; Iizuka, R.; Itoh, A.; et al. The X-ray Telescope onboard Suzaku. *Publ. Astron. Soc. Jpn.* **2007**, *59*, S9–S21.
20. Tanaka, Y.; Moue, H.; Holt, S.S. The X-ray astronomy satellite ASCA. *Publ. Astron. Soc. Jpn.* **1993**, *46*, 37–41.
21. Vernani, D. Advanced manufacturing techniques for X-ray and VHE Gamma-ray astronomical mirrors. Ph.D. Thesis, University of Insubria, Varese, Italy, 2011; pp. 42–55.
22. Navalkar, V.; et al. Pre-flight evaluation of the soft X-ray telescope optics aboard AstroSat, *J. Astrophys. Astron.* **2021**, *42*(2).
23. Weisskopf, M.C.; Ramsey, B.; O'Dell, S.L.; Tennant, A.; Elsner, R.; Soffita, P.; Bellazzini, R.; Costa, E.; Kolodziejczak, J.; Kaspi, V.; et al. The Imaging X-ray Polarimetry Explorer (IXPE). *Results Phys.* **2016**, *6*, 1179–1180.
24. Zhang, S.; Santangelo, A.; Feroci, M.; Xu, Y.; Lu, F.; Chen, Y.; Feng, H.; Zhang, S.; Brandt, S.; Hernanz, M.; et al. The enhanced X-ray Timing and Polarimetry mission—eXTP. *Sci. China Phys. Mech. Astron.* **2019**, *62*, 29502.
25. Zhang, S.N.; Feroci, M.; Santangelo, A.; Dong, Y.W.; Feng, H.; Lu, F.J.; Nandra, K.; Wang, Z.S.; Zhang, S.; Bozzo, E.; et al. eXTP: Enhanced X-ray Timing and Polarization mission. In *Proceedings of the Space Telescopes and Instrumentation 2016: Ultraviolet to Gamma Ray, Edinburgh, UK, 26 June–1 July 2016*; SPIE: Bellingham, WA, USA, 2017; Volume 9905, p. 99051Q.

Chapter 5

Surface modification techniques for microstructural enhancement in metal additive manufacturing

Rupam Tripathi, Jaya Kori, and Somnath Chattopadhyaya

5.1 INTRODUCTION

Metal additive manufacturing (AM) has emerged as a modern manufacturing technology that has enabled the production of complex and tailor-made components with unique design freedom.[1] However, the microstructural integrity of the additively manufactured product is the critical aspect that affects the mechanical properties and performance of the components.[2] Surface mechanical treatments are an effective way to enhance the microstructural characteristics and material properties of AM products. Mechanical surface treatment methods such as shot peening, laser shock peening and ultrasonic impact treatment have the capability of enhancing the surface and near-surface region properties without affecting the material composition (bulk). These treatment methods induce controlled plastic deformation by introducing compressive residual stresses on the surface and sub-surface regions and refining the grain structures, which facilitates enhanced wear, fatigue and corrosion resistance.[3]

The microstructure refers to the size, distribution and arrangement of the grains and phase transformation which influence the mechanical and thermal properties of the manufactured component.[4] The mechanical properties such as strength, toughness, fatigue and ductility are primarily dependent on the microstructures. In the case of metal additive-manufactured products, the rapid thermal and directional solidification rates during layer-by-layer printing lead to the formation of unique microstructural features (grain orientation, interlayer bonding, residual stresses).[5] These features also affect the anisotropy and defect formation in the components. In addition, the microstructural refinement affects the fatigue life and fracture resistance of the additively manufactured products. The fine-grained homogeneous microstructures, along with reduced defect densities, can mitigate the initiation of cracks.[6] Moreover, microstructural analysis techniques such as scanning electron microscopy, electron beam scattered diffraction, transmission electron microscopy, x-ray diffraction, and mechanical testing provide comprehension of grain morphology, crystallographic phases,

DOI: 10.1201/9781003713364-5

residual stresses, and mechanical behaviour, assisting in understanding the material-property correlation. These microstructural characteristics (phase transformation, grain growth kinetics) are also governed by the build parameters such as laser scan speed, laser power, layer thickness and material composition.[7] To achieve the desired microstructural characteristics, an integration of surface treatment methods with optimised build parameters can tailor the mechanical properties and microstructural characteristics of the components as per the requirements.[8] This chapter is an overview of various mechanical surface modification techniques that are employed to refine the microstructures of the AM components, highlighting the significance of material properties in achieving the desired performance of the components.

5.2 MICROSTRUCTURAL CHARACTERISTICS OF METAL AM COMPONENTS

The resulting microstructures in the metal additive manufacturing processes such as powder bed fusion, direct energy deposition and binder jetting directly influence the mechanical properties of the component.[1] Therefore, it is essential to understand the intricate microstructural features and their impact on the material behaviour.

In any AM process, the rapid directional solidification of the metal powders or wire occurs during the melting and layer deposition. The formation of the fine grains in the melt pool and near-surface regions occurs due to the localised melting and cooling.[9] The shape and size of the melt pools are controlled by the process parameters (laser scan speed, power, etc.) that will result in finer grains (for small melt pools). Furthermore, the thermal gradients control the nucleation and grain growth characteristics. The different AM processes are also responsible for the formation of the grain morphology.[10] AM processes such as selective laser melting and electron beam melting upon solidification result in equiaxed grains, and processes such as direct energy deposition result in columnar grains. The thermal gradient and the continuous layer depositions facilitate the formation of such grain structures and influence the mechanical performance of the material.[11] Another important aspect of the grain structures is the effect of the build parameters. The thicker layers due to the longer solidification rates result in the formation of larger grains; however, the thinner layers promote the formation of finer grains.[12] The higher scanning speed and laser power result in the formation of coarse grains with large melt pools. In addition to this, the powder characteristics and alloy composition influence the phase transformation and grain growth phenomenon.[13] Thus, the grain structures and orientation significantly affect the material's performance in terms of strength, fatigue behaviour, microhardness, ductility and anisotropy.

5.3 SURFACE MODIFICATION TECHNIQUES FOR AM COMPONENTS

5.3.1 Shot peening

5.3.1.1 Mechanism

The shot peening (SP) technique is a well-known surface treatment process for aerospace, automobile and biomedical industries that induces significant surface modifications in the components by imposing compressive residual stress in the near-surface regions through the impact of shots (round glass, metallic or ceramic) on the surface.[14] Figure 5.1 depicts the schematic principle of the SP process. It is a cold working process in which high-velocity shots impact the surface, causing plastic deformation on the surface and near-surface regions, resulting in work hardening and refinement of the grains in the near-surface regions.

In the SP process, the impact of each shot impacts localised plastic deformation. The repeated impacts of these shots cause overlapping regions on the surface which are responsible for the modifications in the mechanical properties and microstructural characteristics on the surface and near-surface regions. Upon impact of the shots, the surface undergoes expansion (plastically), which is restrained by the compression of the surrounding materials (Figure 5.1). This introduces compressive residual stresses on the surface and near-surface regions and prevents fatigue failure of the components. The presence of compressive residual stresses also enhances the resistance to stress corrosion cracking. Additionally, during the severe plastic deformation by the impact of shots, the high-strain regions created on

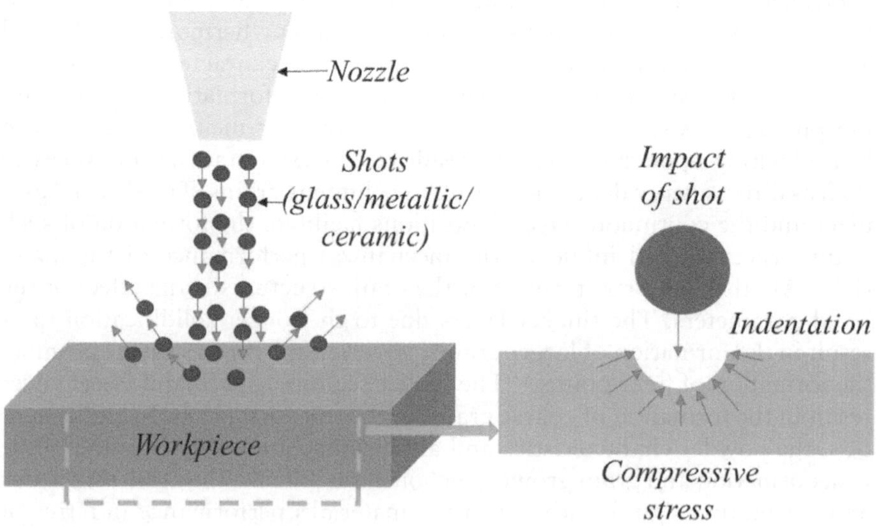

Figure 5.1 Schematic of SP process.

the surface cause the formation of dislocations where the grain rearranges and results in finer grains. This grain refinement or strain hardening resists further deformation as the dislocation density prevents the motion of the dislocations (slips, twinnings), thereby enhancing the strength and hardness on the surface and sub-surface regions. The formation of finer grains provides a barrier to crack propagation and micro-crack generation and thereby enhances the life of the component.

5.3.1.2 Microstructural modification in AM components by SP

Using SP, AlMangour and Yang[15] attempted to improve the mechanical properties and surface quality of 17-4 stainless steel components manufactured by direct metal laser sintering. SP on the as-built component of 17-4 stainless steel resulted in severe plastic deformation on the surface layers causing grain refinements, thereby enhancing the mechanical properties. The study showed significant enhancement in the compressive residual stress in the peened samples with favourable surface roughness, microhardness and wear resistance due to the martensitic phase transformation of restrained austenite in as-built samples. Similarly, Zhang et al.[16] investigated surface modification using high-energy SP (HESP) to achieve nanocrystallisation modification in SLS-manufactured stainless-steel 316L parts. The microstructural study using optical microscopy and transmission electron microscopy (TEM) showed that the HESP treatment resulted in grain refinement with an average grain size of 25 nm. This surface nanocrystallisation evolution resulted from the interactions of the high-density dislocations, tangles and twins (Figure 5.2).

Suwanpreecha et al.[17] explored the effect of ageing and SP (individually and in combination) on the mechanical properties, surface integrity and microstructure in comparison to the untreated material extrusion additively manufactured 17-4PH stainless steel. The microstructural analysis depicted the presence of martensitic structures along with the δ-ferrite on the grain boundaries with the presence of inclusions and porosity in all the samples. The peened samples showed work hardening up to a depth of 140 μm with nanocrystalline surfaces (100–500 nm) and compressive residual stresses up to a range of 600 MPa (approximately). Walczak and Szala[18] also conducted the surface property, wear and corrosion analysis of the 17-4PH stainless steel additively manufactured by direct metal laser sintering. The study revealed that the SP caused microstructural grain refinement (up to 14 nm) inducing martensite formation and reducing austenite formation. Additionally, the peening samples showed an increase in surface hardening (from 247 HV to 542 HV). Similarly, the multi-layer deposition by wire arc additive manufacturing (WAAM) along with plasma spray coating (Ni58Cr20Fe5Mo10) of duplex stainless steel 2205 using 23Cr9Ni4Mo2Mn (FeAs) filler has shown improved microstructures and

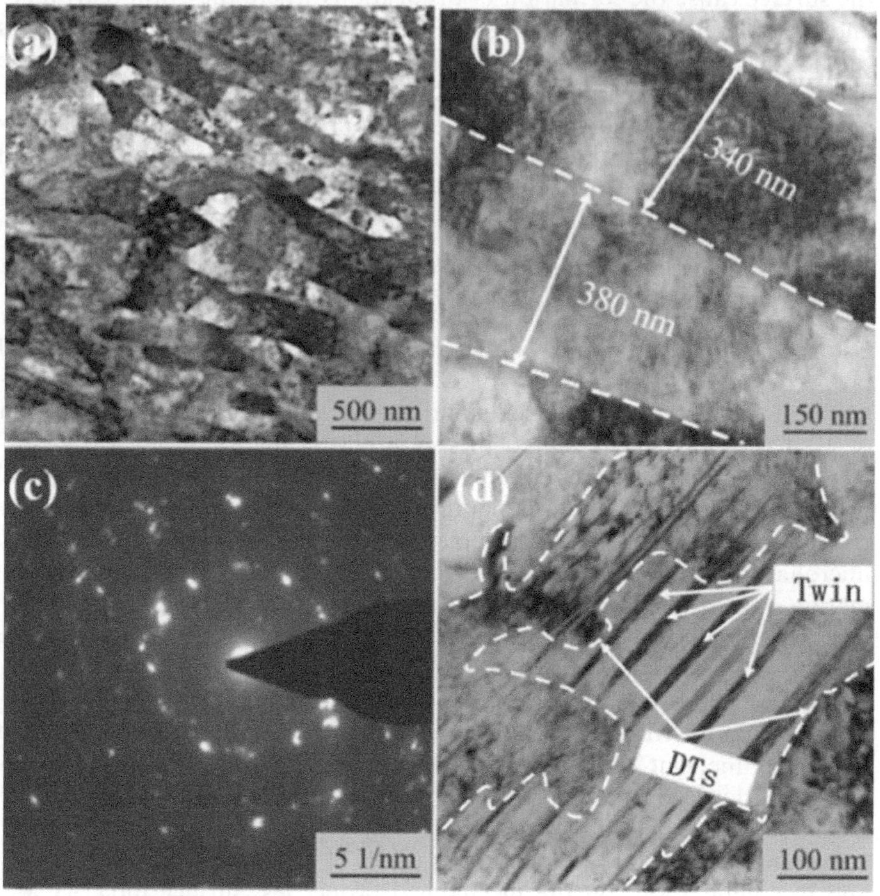

Figure 5.2 (a) TEM of ultrafine grain, (b) detailed elongated ultrafine grain, (c) electron diffraction pattern, (d) twin and dislocation tangles interaction. (From Zhang et al., *J Mater Res Tech*, 30:4322–4329.)

corrosion resistance. The microstructural analysis revealed the presence of different types of structures such as solid-phase martensite, acicular austenite and martensite–austenite phases.[19]

5.3.1.3 Advantages and limitations of the SP

The SP technique, which is employed for enhancing the microstructure characteristics and mechanical properties of the metal AM components, has advantages such as (i) improving the fatigue life of the components which is beneficial for structural, aerospace and automotive applications subjected to cyclical loading conditions; (ii) mitigating the tensile residual stresses which are generated during the manufacturing of AM components due to the directional solidification of the layers, which reduces stress corrosion cracking and deformation; (iii) increasing the hardness of the surface and near-surface regions due to the work-hardening effect, thereby, improving wear resistance and surface damage. However, there are certain disadvantages associated with the techniques: (i) uniform coverage in complex or intricate surfaces (complex lattice structures) is challenging, (ii) the increase in the surface roughness caused by the impact of the shots is another crucial factor in balancing between surface enhancement and surface integrity and (iii) the depth of the enhancement is limited in this technique, therefore to have a deeper depth of enhancement multiple passes may be needed.[20]

5.3.2 Laser shock peening

5.3.2.1 Mechanism

Laser shock peening is a surface modification technique which overcomes the limitations of conventional shot peening by offering deeper penetration of plastic deformation. It involves a high-density pulsed laser beam which strikes the material surface and generates a plasma shock wave on the surface (Figure 5.3). The treated surface is initially covered with a thin absorbing layer made by paint, plastic tape or any other transparent medium. These shock waves when propagated through the surface of the component cause plastic deformation which results in the generation of compressive residual stress on the surface.

Owing to the presence of the compressive residual stresses, the fatigue life, hardness and yield strength of the material are enhanced remarkably. This increase is attributed to the high dislocation density, phase transformations and presence of dislocations such as twins caused by the propagation of the shock wave.[21] Additionally, the wear resistance and corrosion resistance of the component are also enhanced. LSP is accompanied by significant changes in the microstructures and phases. These changes are observed through SEM, XRD and TEM. These microstructural changes are influenced by the laser parameters and treatment conditions. Various investigations using LSP on marginal steels have shown the phase transformation from austenite to martensite and a significant enhancement in the dislocation density.[22] Additionally, the presence of α-phase along the twin interaction was also reported in austenitic stainless steel 304.

5.3.2.2 Microstructural modification in AM components by LSP

In the case of AM components, it was observed that the high-performance integrated SLM along with LSP has resulted in enhanced mechanical properties and refined grain structures in the surface layers of Ti6Al4V alloy components.[23] This integrated approach has resulted in high-density

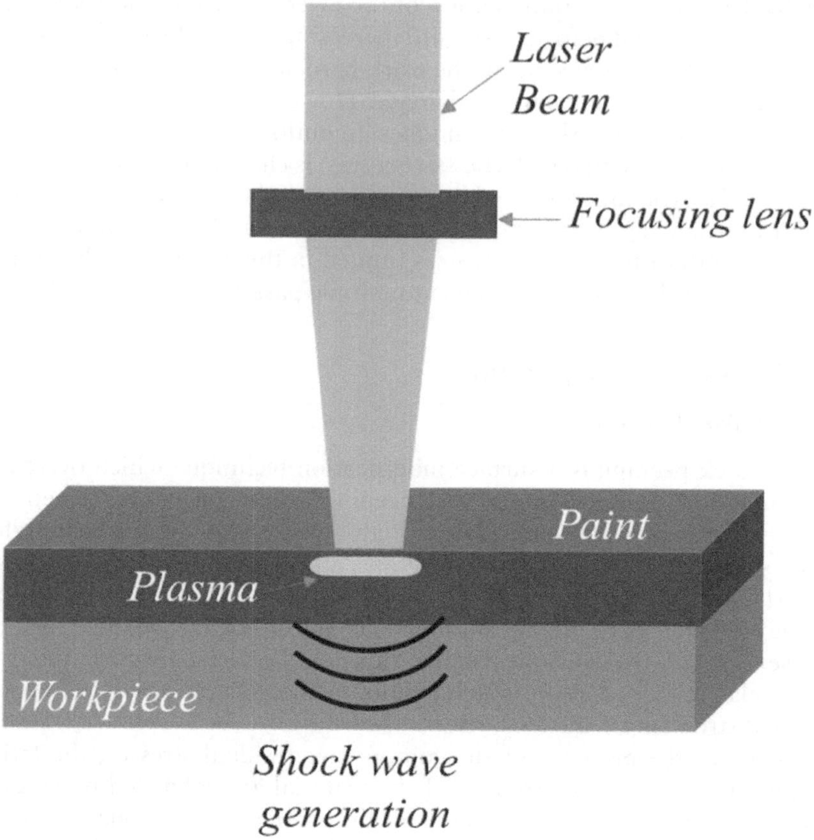

Figure 5.3 Schematic showing LSP.

dislocations and mechanical twins in the coarse α′ martensites by the propagation of the laser shock waves which evolved into refined α′ martensites (Figure 5.4).

Maleki et al.[24] investigated the effect of LSP on the corrosion resistance of laser power bed fused (L-PBF) AlSi10Mg samples. The study focused on determining the effect of laser pulse overlaps and energies on microstructure, surface texture, porosity, hardness, residual stresses and corrosion resistance. The results revealed that the laser overlaps and energies have induced high plastic strains which resulted in higher grain refinement along with improved surface roughness and hardness in the sub-surface regions up to a depth of 430 μm.

Singh and Deoghare[25] demonstrated the microstructural effect of LSP on the AM Ti6Al4V sample with a thick layer deposited by the laser direct energy deposition technique. It was observed that the deposition was formed without major defects such as microcracks that are evident

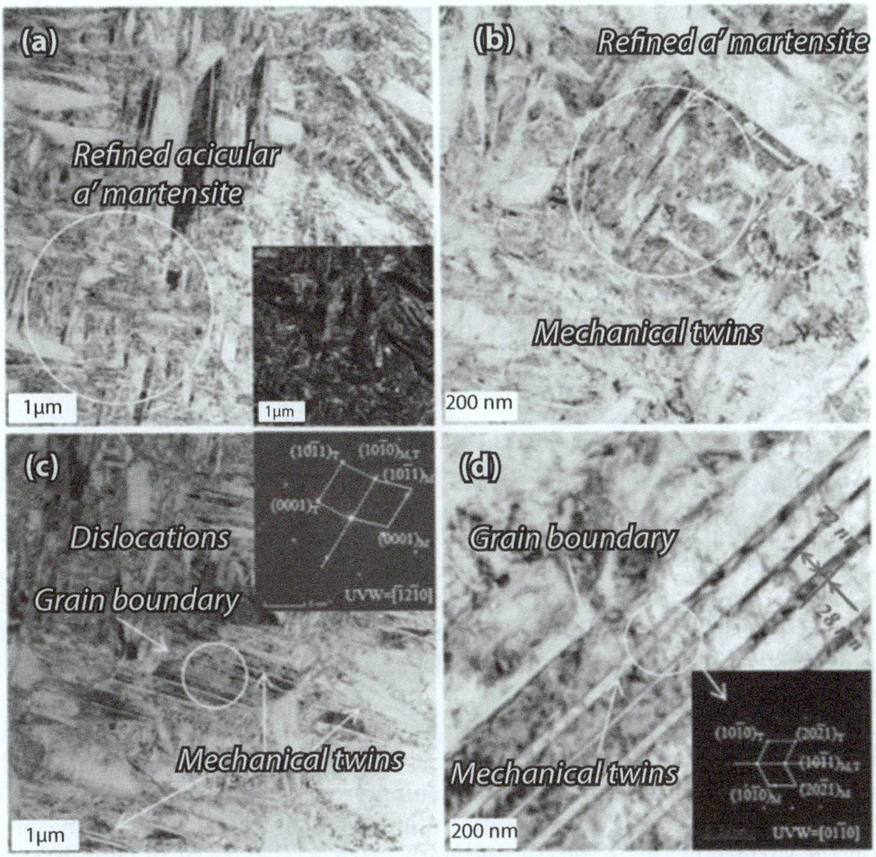

Figure 5.4 TEM image of SLM-LSP samples showing (a) acicular α′ martensite, (b) refined α′ martensite and mechanical twins, (c) dislocations and mechanical twins and (d) mechanical twins. (From Lu et al., 2020, *Int J Mach Tools Manuf.* 148.)

in the AM samples manufactured with thicker layers. The presence of microcracks was observed only on the surface of the sample along with the Widmanstatten pattern. Moreover, the LSP also enhanced the microhardness by 49% and fatigue performance by 40% owing to the grain refinement and strain hardening.[25] Similarly, S. Kumar et al.[26] examined the integration of the LSP and heat treatment on the SLM-manufactured 17-4PH stainless steel. It was observed that the LSP has resulted in the formation of microsized grain refinement along with the accumulation of the dislocations which was responsible for the enhanced mechanical properties (residual compressive strength = 850 MPa approximately). Additionally, high misorientation angles and high-angle grain boundaries were observed in LSP and heat-treated AM samples which affect the microhardness in the near-surface regions.

5.3.2.3 *Advantages and limitations of LSP*

The microstructural evolution by LSP in AM components has the following advantages: (i) it doesn't require any physical connection with the tool thereby making the techniques favourable for complex geometries, (ii) it offers deeper penetration/hardening depth as compared to conventional surface modification techniques and (iii) precise control over parameters makes it an effective surface treatment process. Although this technique has shown promising advantages it has certain limitations such as (i) high initial investment, (ii) complexity in optimising the peening parameters, (iii) the LSP process is comparatively slower than the other peening methods, and (iv) the laser energy can cause localised melting of the component, thereby affecting its surface integrity.[27,28]

5.3.3 Ultrasonic impact treatment

5.3.3.1 *Mechanism*

Ultrasonic impact treatment (UIT), also known as high-frequency mechanical treatment (HFMT), is the high-frequency surface strengthening process which uses high-impact energy to create large plastic deformations on the material surface. This high-impact energy is driven by ultrasound (high power: 20–55 kHz) through a transducer which converts electrical energy into high-frequency vibration (20–50 μm) which is transferred to the needle attached to the transducer (Figure 5.5). This high-frequency amplitude of the needle impacts the material surface (vibrations converted to mechanical pulses) thereby initiating the strengthening process.[29]

The plastic deformation is initially imparted by the impact of the pin on the surface of the material, and with further increase in the number of impacts, the severity of the plastic deformation increases resulting in the work hardening of the surface. The microstructural modifications by UIT are responsible for the enhanced mechanical properties. The grain refinement and phase transformation facilitate the increase in hardness, wear resistance and corrosion in the components. Moreover, the process parameters such as impact angle, coverage, frequency, number of impacts, velocity, pin or needle shape, diameter and length are also responsible for the enhancement in the mechanical properties (fatigue strength, yield strength, microhardness) and resistance to wear.[29]

5.3.3.2 *Microstructural modification in AM components by UIT*

UIT is preferred to mitigate the defects encountered in wire arc additive manufacturing (WAAM) components. Wang et al.[30] examined the effect of UIT on the mechanical properties and microstructure characteristics of WAAM 2219 aluminium alloy. The microstructural characteristics (grain

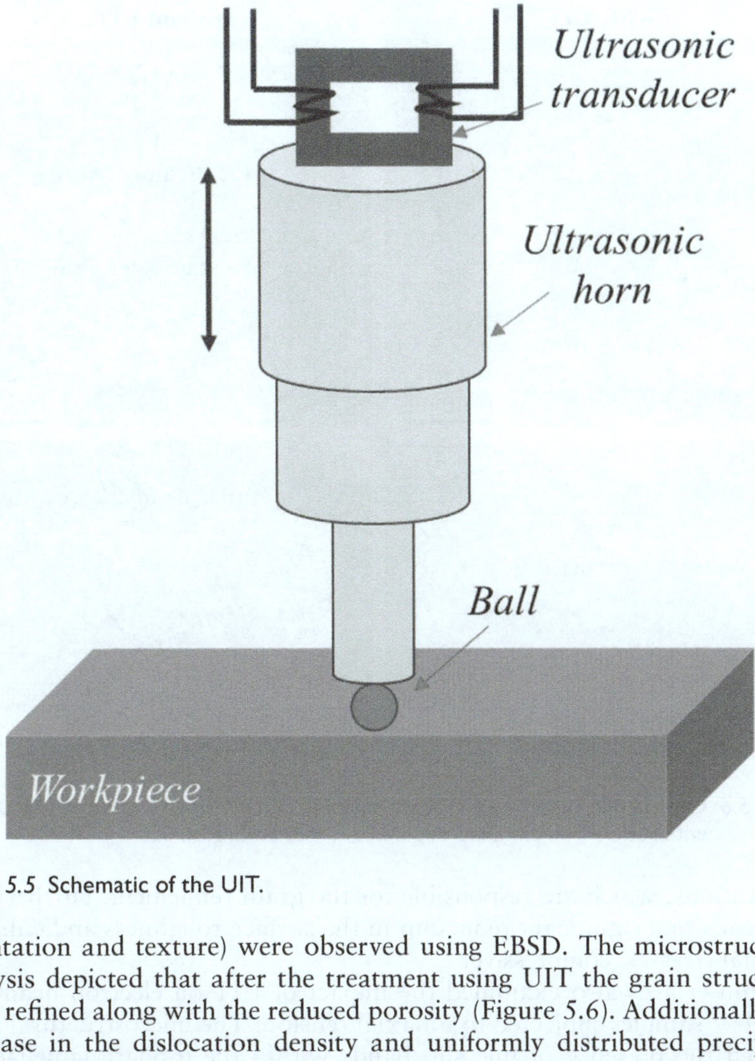

Figure 5.5 Schematic of the UIT.

orientation and texture) were observed using EBSD. The microstructural analysis depicted that after the treatment using UIT the grain structures were refined along with the reduced porosity (Figure 5.6). Additionally, the increase in the dislocation density and uniformly distributed precipitate phase were observed.

In the case of ER321 stainless steel fabricated by WAAM-assisted UIT, samples[31] showed the change in the grain structure from coarse columnar dendrites to fine columnar dendrites with an effective grain size reduction of ~150%. UIT has also facilitated recrystallisation in the samples which reduced the texture and dislocation densities. This enhancement in the grain structure leads to an increase in the yield strength (~10%) and microhardness (~12%).

Lv et al.[32] investigated the microstructural evolution of the Ti6Al4V WAAM samples treated with double-sided UIT. This double-sided treatment showed the stacking fault structures along with high-density

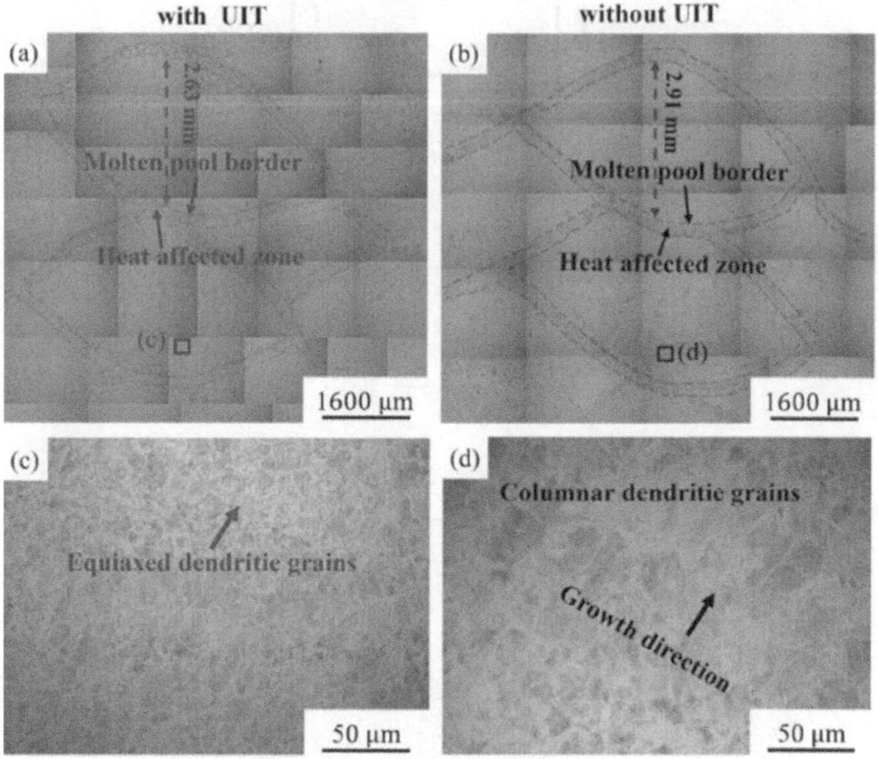

Figure 5.6 Optical micrographs of WAAM samples (a, c) with UIT treatment and (b, d) without UIT. (From Diao et al., 2022, *Mat Sci Eng: A*, 857

dislocations, which are responsible for the grain refinement. Furthermore, this leads to a significant reduction in the surface roughness and enhances residual stresses (compressive).

Panin et al.[33] also examined the impact of UIT on electron beam AM Ti6Al4V samples subjected to uniaxial tension. The microstructural study showed the presence of fine slip bands within the α-phase lamellae and coarse slip bands within packet martensite, hindering the dislocation sliding. This intensive deformation leads to the formation of ductile microrelief on the surface.

5.3.3.3 Advantages and limitations of ultrasonic impact treatment (UIT)

UIT has distinct advantages that cause the effectiveness in increasing the mechanical properties and grain refinement of the AM components: (i) the moderate depth penetration minimises the surface damage, thereby advantageous for delicate and intricate components, (ii) precise control of the parameters helps to tailor the treatment process for different types of materials and geometry, (iii) the induced compressive residual stress causes significant enhancement in the fatigue life of the component. However, as compared to other surface modification techniques it has certain limitations: (i) the UIT strengthening depth is limited, (ii) reaching the intricate sections of the components is difficult and (iii) there is a high setup cost with skilled operators, and maintenance is needed.[30]

5.4 CONCLUSION AND FUTURE SCOPE

Mechanical surface treatment methods have the potential to significantly enhance the microstructural and mechanical properties in metal additive manufactured components. The surface mechanical treatment techniques (SP, LSP and UIT) addressed in this study demonstrated their application in mitigating the inherent defects in AM parts by effectively increasing the dislocation densities, refining the grains and inducing compressive residual stresses. These microstructural evolutions significantly impact the mechanical properties (fatigue strength, hardness, corrosion and wear resistance) and can successfully optimise the parameters for specific applications (aerospace, automobiles, biomedical) and geometries. However, in the future, the effect of integrated surface mechanical techniques on the performance of the AM components can be studied.[24] Additionally, the combined effect of these surface mechanical treatments along with other thermal treatments and coatings can be explored to address the challenges in AM components.[26]

REFERENCES

1. Durai Murugan P, Vijayananth S, Natarajan MP, Jayabalakrishnan D, Arul K, Jayaseelan V, Elanchezhian J. 2022. A current state of metal additive manufacturing methods: A review. *Mater Today Proc.* 59:1277–1283. doi:10.1016/j.matpr.2021.11.503
2. Armstrong M, Mehrabi H, Naveed N. 2022. An overview of modern metal additive manufacturing technology. *J Manuf Process.* 84:1001–1029. doi:10.1016/j.jmapro.2022.10.060
3. Kumar D, Idapalapati S, Wang W, Narasimalu S. 2019. Effect of surface mechanical treatments on the microstructure-property-performance of engineering alloys. *Materials.* 12(16). doi:10.3390/ma12162503
4. Peyre P, Scherpereel X, Berthe L, Carboni C, Fabbro R, Be G. 2000. Surface modifications induced in 316L steel by laser peening and shot-peening. Influence on pitting corrosion resistance. *Mater Sci Eng: A.* 280:294–302.
5. Liu Z, Zhao D, Wang P, Yan M, Yang C, Chen Z, Lu J, Lu Z. 2022. Additive manufacturing of metals: Microstructure evolution and multistage control. *J Mater Sci Technol.* 100:224–236. doi:10.1016/j.jmst.2021.06.011

6. Elangeswaran C, Cutolo A, Muralidharan GK, Vanmeensel K, Van Hooreweder B. 2020. Microstructural analysis and fatigue crack initiation modelling of additively manufactured 316L after different heat treatments. *Mater Des*. 194. doi:10.1016/j.matdes.2020.108962
7. Ahmed N, Barsoum I, Haidemenopoulos G, Al-Rub RKA. 2022. Process parameter selection and optimization of laser powder bed fusion for 316L stainless steel: A review. *J Manuf Process*. 75:415–434. doi:10.1016/j.jmapro.2021.12.064
8. Santos V, Uddin M, Hall C. 2023. Mechanical surface treatments for controlling surface integrity and corrosion resistance of Mg alloy implants: A review. *J Funct Biomater*. 14(5). doi:10.3390/jfb14050242
9. Jafari D, Vaneker THJ, Gibson I. 2021. Wire and arc additive manufacturing: Opportunities and challenges to control the quality and accuracy of manufactured parts. *Mater Des*. 202. doi:10.1016/j.matdes.2021.109471
10. Roehling TT, Shi R, Khairallah SA, Roehling JD, Guss GM, McKeown JT, Matthews MJ. 2020. Controlling grain nucleation and morphology by laser beam shaping in metal additive manufacturing. *Mater Des*. 195. doi:10.1016/j.matdes.2020.109071
11. Meng X, Min J, Sun Z, Zhang W, Chang H, Han Y. 2021. Columnar to equiaxed grain transition of laser deposited Ti6Al4V using nano-sized B4C particles. *Compos B Eng*. 212. doi:10.1016/j.compositesb.2021.108667
12. Wrobel R, Del Guidice L, Scheel P, Abando N, Maeder X, Vassiliou M, Hosseini E, Spolenak R, Leinenbach C. 2024. Influence of wall thickness on microstructure and mechanical properties of thin-walled 316L stainless steel produced by laser powder bed fusion. *Mater Des*. 238. doi:10.1016/j.matdes.2024.112652
13. Agarwal N, Monu MCC, Tamil Selvam K, Obeidi MA, Brabazon D. 2023. Study of the effects of laser power and scanning speed on the microstructural morphologies and physical properties of L-PBF produced Ni52.39Ti47.61. *J Mater Res Technol*. 27:8334–8343. doi:10.1016/j.jmrt.2023.11.146
14. Scholtes B, Vöhringer O. 2016. Mechanical surface treatment. In: *Reference Module in Materials Science and Materials Engineering* [Internet]. Elsevier. doi:10.1016/B978-0-12-803581-8.03549-9
15. AlMangour B, Yang JM. 2016. Improving the surface quality and mechanical properties by shot-peening of 17-4 stainless steel fabricated by additive manufacturing. *Mater Des*. 110:914–924. doi:10.1016/j.matdes.2016.08.037
16. Zhang C, Wu J, Xu R, Zhu Y, Xu B, Yang H. 2024. Surface nanocrystallization mechanism of additive manufactured 316L stainless steel via high-energy shot peening. *J Mater Res Technol*. 30:4322–4329. doi:10.1016/j.jmrt.2024.04.178
17. Suwanpreecha C, Linjee S, Newyawong P, Yordsri V, Songkuea S, Wutikhun T, Manonukul A. 2024. Effects of aging and shot peening on surface quality and fatigue properties of material extrusion additive manufactured 17-4PH stainless steel. *Mater Des*. 241. doi:10.1016/j.matdes.2024.112939
18. Walczak M, Szala M. 2021. Effect of shot peening on the surface properties, corrosion and wear performance of 17-4PH steel produced by DMLS additive manufacturing. *Arch Civ Mech Eng*. 21(4). doi:10.1007/s43452-021-00306-3

19. Selvabharathi R. 2023. Microstructure and corrosion resistance of 23Cr9Ni4Mo2Mn filler duplex stainless steel 2205 fabricated by wire arc additive manufacturing with subsequent severe shot peening and Ni58Cr20Fe5Mo10 plasma spray coating. *Mater Today Commun*. 37:107560. doi:10.1016/j.mtcomm.2023.107560
20. Zhang X, Huang S, Li D, Geng J, Yang F, Li Q. 2022. An approach to improve the microstructure and mechanical properties: A hybrid manufacturing of laser directed energy deposition and shot peening. *Addit Manuf* [Internet]. 55:102686. doi:10.1016/j.addma.2022.102686
21. Ding K, Ye L. 2006. 1 – General introduction. In: Ding K, Ye L, editors. *Laser Shock Peening* [Internet]. Woodhead Publishing; p. 1–6. doi:10.1533/9781845691097.1
22. Ding K, Ye L. 2006. Laser shock peening performance and process simulation. Boca Raton, CRC Press, 2006. 162. doi:10.1201/9781439823620
23. Lu J, Lu H, Xu X, Yao J, Cai J, Luo K. 2020. High-performance integrated additive manufacturing with laser shock peening–induced microstructural evolution and improvement in mechanical properties of Ti6Al4V alloy components. *Int J Mach Tools Manuf.* 148. doi:10.1016/j.ijmachtools.2019.103475
24. Maleki E, Unal O, Shao S, Shamsaei N. 2023. Effects of laser shock peening on corrosion resistance of additive manufactured AlSi10Mg. *Coatings*. 13(5). doi:10.3390/coatings13050874
25. Singh SN, Deoghare AB. 2024. Microstructure, microhardness, tensile and fatigue investigation on laser shock peened Ti6Al4V manufactured by high layer thickness directed energy deposition additive manufacturing. *Opt Laser Technol* [Internet]. 177:111132. doi:10.1016/j.optlastec.2024.111132
26. S PK, P DB, Gautam J, Rai AK, Paul CP. 2024. Synergistic integration of laser shock peening and heat treatment for refined microstructure and enhanced mechanical properties in additively manufactured 17–4PH stainless steel. *J Mater Process Technol* [Internet]. 328:118395. doi:10.1016/j.jmatprotec.2024.118395
27. Deng W, Wang C, Lu H, Meng X, Wang Z, Lv J, Luo K, Lu J. 2023. Progressive developments, challenges and future trends in laser shock peening of metallic materials and alloys: A comprehensive review. *Int J Mach Tools Manuf* [Internet]. 191:104061. doi:10.1016/j.ijmachtools.2023.104061
28. Jiang Q, Li S, Zhou C, Zhang B, Zhang Y. 2021. Effects of laser shock peening on the ultra-high cycle fatigue performance of additively manufactured Ti6Al4V alloy. *Opt Laser Technol* [Internet]. 144:107391. doi:10.1016/j.optlastec.2021.107391
29. Sun L, Huang L, Wu P, Huang R, Fang N, Xu F, Xu K. 2023. Progress on the effect and mechanism of ultrasonic impact treatment on additive manufactured metal fabrications. *Crystals* (Basel). 13(7). doi:10.3390/cryst13070995
30. Wang C, Li Y, Tian W, Hu J, Li B, Li P, Liao W. 2022. Influence of ultrasonic impact treatment and working current on microstructure and mechanical properties of 2219 aluminium alloy wire arc additive manufacturing parts. *J Mater Res Technol*. 21:781–797. doi:10.1016/j.jmrt.2022.09.055
31. Diao M, Guo C, Sun Q, Jiang F, Li L, Li J, Xu D, Liu C, Song H. 2022. Improving mechanical properties of austenitic stainless steel by the grain refinement in wire and arc additive manufacturing assisted with ultrasonic impact treatment. *Mater Sci Eng: A*. 857. doi:10.1016/j.msea.2022.144044

32. Lv J, Alexandrov I V, Luo K, Lu H, Lu J. 2022. Microstructural evolution and anisotropic regulation in tensile property of cold metal transfer additive manufactured Ti6Al4V alloys via ultrasonic impact treatment. *Mater Sci Eng: A* [Internet]. 859:144177. doi:10.1016/j.msea.2022.144177
33. Panin AV, Kazachenok MS, Dmitriev AI, Nikonov AY, Perevalova OB, Kazantseva LA, Sinyakova EA, Martynov SA. 2022. The effect of ultrasonic impact treatment on deformation and fracture of electron beam additive manufactured Ti-6Al-4V under uniaxial tension. *Mater Sci Eng: A* [Internet]. 832:142458. doi:10.1016/j.msea.2021.142458

Chapter 6

Viscous fluid-enhanced tribological performance on variable load and groove-textured cylinder systems

Krishnadas Narayanan Nampoothiri and Shubrajit Bhaumik

6.1 INTRODUCTION

Depletion of non-renewable sources and an increase in demand for tools in various automobile and other mechanical sectors have led to the importance of improvement in tribological performance of moving contact and non-contact systems (Teh et al. 2022; Tung and McMillan 2004; Ciulli 2019). Enhanced tribological performance reduces the concerns of wear, friction and lubrication (Bhaumik et al. 2022; Haribabu et al. 2023). Thus, it is essential to improve the tribological performance of mechanical systems like cutting tools (Astakhov 2013), piston rings and cylinder liners (Lu and Wood 2020), and journal bearings (Bompos and Nikolakopoulos 2016). One way of improving the tribological performance is to introduce surface texturing on the moving systems (Vishnoi, Kumar, and Murtaza 2021). Surface textures created on the surfaces of systems act as storage reservoirs for lubricants, thus ensuring an improvement in the lubrication performance (Yin et al. 2023; Ghani et al. 2021; Lu and Wood 2020).

Various forms of surface texturing are implemented on the systems using different fabrication techniques. The most common surface textures are dimples (Ali et al. 2023), grooves (Fan et al. 2019), pillars (Ammosova, Mönkkönen, and Suvanto 2020) and inverted pyramids (Basher et al. 2019). These structures are achieved either by laser machining techniques (Pietrzyk et al. 2019), injection moulding methods (Ammosova, Mönkkönen, and Suvanto 2020) or by other means such as chemical etching (Basher et al. 2019) and milling processes (Jain and Bajpai 2019).

Among the most researched surface texturings, groove texturing is the popular surface texturing geometry due to its ease of fabrication and simplicity (Shen et al. 2021; Wu et al. 2016; Bhaumik et al. 2022). Mao et al. (2020) fabricated grooves of width 30 μm and depth 12 μm using a laser patterning technique. Shen et al. (2021) investigated the frictional performance of certain specific materials with micro-grooves fabricated on the metal bearings for bioimplants.

Even though extensive research on groove-textured geometries has been conducted and compared with non-textured systems, there is a gap in the

DOI: 10.1201/9781003713364-6

literature regarding the optimisation of the groove depth as well as the dependence of various parameters like stress and pressure on load and viscosity. This gap is mainly due to several experimental constraints. Thus, it is important to build a model and investigate these effects through numerical studies. In this study, we simulate and analyse the effect of groove-sized texturing and compare its effect on tribological performance, which is mainly the stress distribution and the lubricant pressure with the non-textured system. The effect of groove depth is also investigated, and we have estimated an optimal groove depth that could significantly improve the tribological performance of the groove-textured rotating cylinder when it is subjected to bulk applied with a certain lubricant viscosity.

6.2 CONSIDERED GEOMETRIES

To investigate the tribological performances, two cylindrical geometries of cylindrical radius 30 mm and height 50 mm are considered. Figure 6.1 represents the schematic diagrams of the two cylinders. Figure 6.1a symbolises the non-textured cylinder. Figure 6.1b denotes the groove-textured cylinder where 11 grooves of 1 mm width, 40 mm length and varying depth (20 μm to 100 μm) are constructed with equal groove spacing of 4 mm. Varying loads of 20 N to 200 N are applied as a normal bulk load to the cylinder as

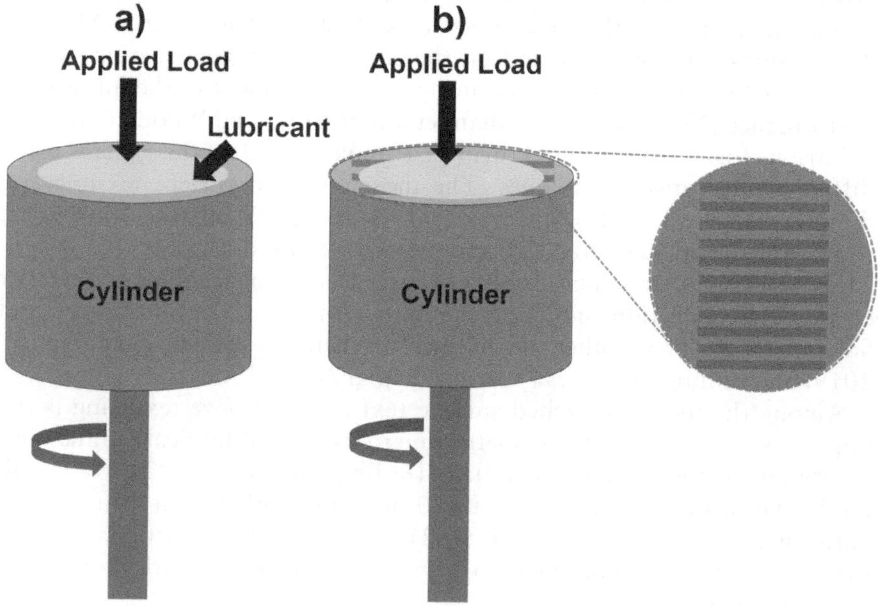

Figure 6.1 Schematic diagrams of the geometries considered for simulations: (a) non-textured cylinder, (b) groove-textured cylinder.

shown in Figure 6.1. Lubricants of varying viscosities (μ = 0.05 to 0.5 Pa·s) are chosen, and these lubricants fill the clearance between the applied load and the cylinder. The cylinder is rotating at different speeds in the range of 30 to 300 rpm.

6.3 MATERIALS

Steel is chosen as the material for the cylinder. The material properties are taken from the literature and are provided as follows: Young's modulus E = 200 GPa, Poisson's ratio υ = 0.28 and density ρ = 8650 kg/m^3. SAE lubricants of different viscosities are chosen. The density of those lubricants is kept constant at ρ = 860 kg/m^3.

6.4 SIMULATION MODEL

The 3D simulation studies regarding the tribological performance of the geometries are investigated using COMSOL Multiphysics® software. For analysis, thin-film flow and solid mechanics are used as the solvers. For analysing the lubricant pressure, the thin-film flow solver is used. The pressure in the lubricant is governed by the Reynolds equation. For an incompressible fluid with no slip condition, the Reynolds equation is given by

$$\nabla_T \cdot \left(\frac{-\rho h^3}{12\mu} \nabla_T p + \frac{\rho h}{2} (v_a + v_b) \right) - \rho \big((\nabla_T b.v_b) - (\nabla_T a.v_a) \big) = 0 \tag{6.1}$$

where h is the lubricant thickness, ρ is the pressure, a is the location of the cylinder base, υ_a is the tangential velocity of the cylinder base, b is the location of the solid wall and υ_b is the tangential velocity of the solid wall. Here we consider the pressure to be constant throughout the lubricant film thickness. Thus, the tangential projection of the gradient operator, $\rho\nabla T$, is used to calculate the pressure distribution on the lubricant surface and the term $\rho\big((\nabla_T b.v_b) - (\nabla_T a.v_a)\big)$ becomes zero (Equation 6.1).

The lubricant thickness h is calculated as

$$h = c(1 + \varepsilon coscos\theta) \tag{6.2}$$

where c is the clearance between the applied load and the cylinder and is taken as 3 mm. ε denotes the eccentricity and is taken as 0.6. θ represents the polar angular coordinate of a point on the lubricant. While u_a refers to the tangential velocity of the cylinder base in the x-axis, υ_a denotes the tangential velocity of the cylinder base in the y-axis. They are calculated as follows:

$$u_a = -\omega R sinsin\theta \quad (6.3)$$

$$v_a = \omega R coscos\theta \quad (6.4)$$

where ω denotes the angular velocity of the cylinder. υ_b is assumed to be zero, and the no-slip wall condition is taken for the film flow model.

For the solid mechanics model, the boundary load is applied to the cylindrical radius as seen in Figure 6.1. All other boundaries are considered as fixed constraints. Steel is considered isotropic in nature, and the values of E and υ, mentioned in Section 6.3, are provided in the solver. Here, a stationary study is considered. The lubricant pressure distribution as well as the stress distribution on the cylinder during loading are solved iteratively using the thin-film and solid mechanics solvers.

6.5 RESULTS AND DISCUSSION

6.5.1 Non-textured cylinder

First, the tribological analysis on the non-textured cylinder (Figure 6.2a) for various parameters, such as applied load, lubricant viscosity and rotational speed of the cylinder, is done using the aforementioned two solvers. Initially, the applied load is kept constant at 120 N. The lubricant viscosity is taken as 0.1 Pa·s and the cylinder is rotating at 150 rpm.

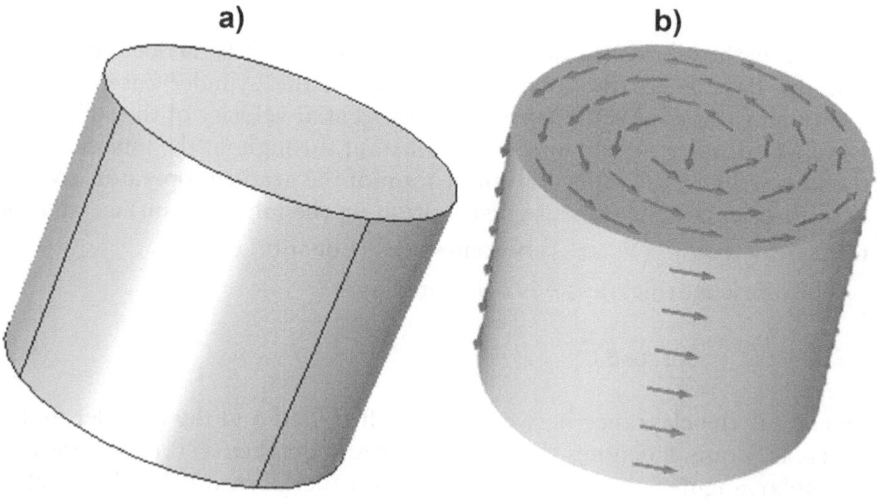

Figure 6.2 (a) Geometry of non-textured cylinder. (b) Geometry showing the base velocity direction with arrows.

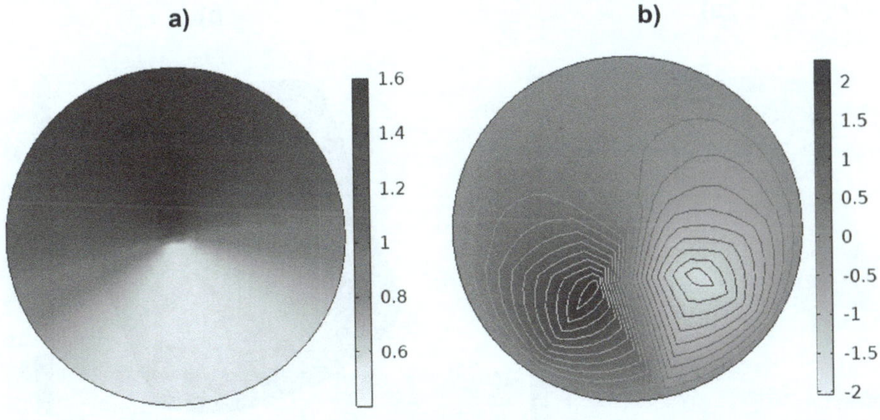

Figure 6.3 Top view contour representation of (a) the lubricant thickness (mm) and (b) lubricant pressure distribution (MPa).

Figure 6.2b shows the base velocity direction of the non-textured cylinder. Here the cylinder is considered as the solid wall, and on this rotating cylinder the Reynold's equation (Equation 6.1) is solved. Figure 6.3a shows the converging and diverging lubricant thickness on the top surface of the cylinder where load is applied. The film thickness spreads non-uniformly due to the lubricant viscosity and rotation of the cylinder. The clearance c also plays a pivotal role in determining the lubrication thickness. In this study, for simplicity, we have kept the magnitude of the clearance constant. The maximum thickness for a viscous fluid of 0.1 Pa·s is estimated to be 1.6 mm. Figure 6.3b shows the pressure distribution of the top cylindrical surface. We could observe that the maximum pressure is attained towards the region where the lubricant thickness is minimum.

Since Reynold's equation does not reveal a direct interpretation of applied load to the structure, we have analysed the stress distribution using the solid mechanics solver. Figure 6.4 depicts the normal stress and the shear stress distribution of the rotating cylinder when it is subjected to 120 N bulk load. From Figure 6.4, maximum stress distribution is felt in the regions where the load is concentrated.

To investigate the tribological performance of the rotating cylinder when it is subjected to a specific load, the effect of stresses and lubrication pressure on certain parameters has been simulated and plotted. Figure 6.5 signifies the relation of maximum normal and shear stresses on the rotating cylinder with applied load. The magnitude of the load is varied from 20 N to 200 N. We could observe linear relations for load with normal stress (Figure 6.5a) and shear stress (Figure 6.5b). The maximum values are felt in the regions where the load is applied. It is important to note that during these studies, other parameters such as lubrication viscosity and cylinder rotation speed are kept constant at 0.1 Pa·s and 150 rpm, respectively.

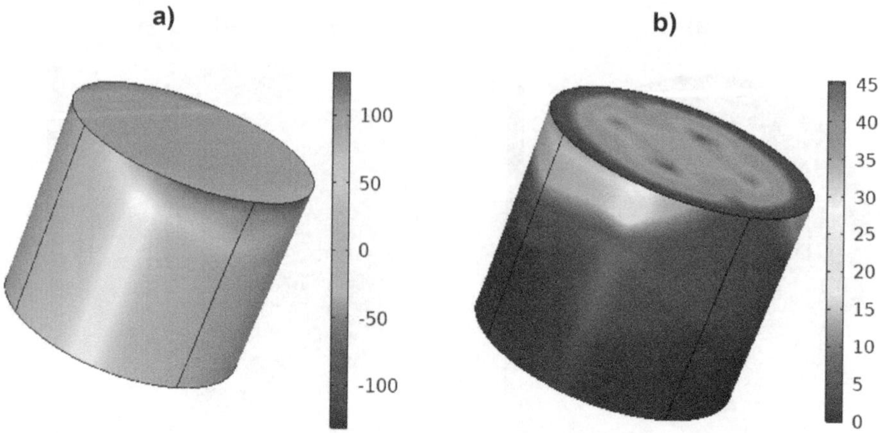

Figure 6.4 Distribution of (a) normal stress and (b) shear stress (in kPa) for na on-textured cylinder when it is subjected to 120 N applied load with lubricant viscosity of 0.1 Pa·s. The cylinder is rotating at 150 rpm.

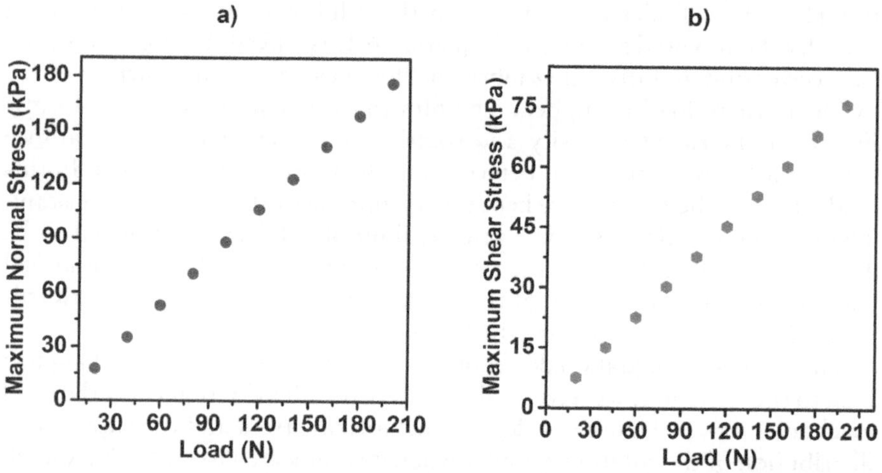

Figure 6.5 Graphs representing the variation of (a) maximum normal stress and (b) maximum shear stress with applied load.

Figure 6.6 depicts the relation of lubricant pressure with lubricant viscosity and cylinder rotational speed. The lubricant pressure is observed to increase with lubricant viscosity (Figure 6.6a).

The viscosity chosen is in the range 0.05 to 0.5 Pa·s, which represents the viscosities of the lubricants available and used for experiments. Here the load and the cylinder rotation are kept constant at 120 N and 150 rpm, respectively. A similar trend is also observed for lubricant pressure with the rotational speed of the cylinder for an applied load at 150 rpm and lubricant viscosity at 0.1 Pa·s.

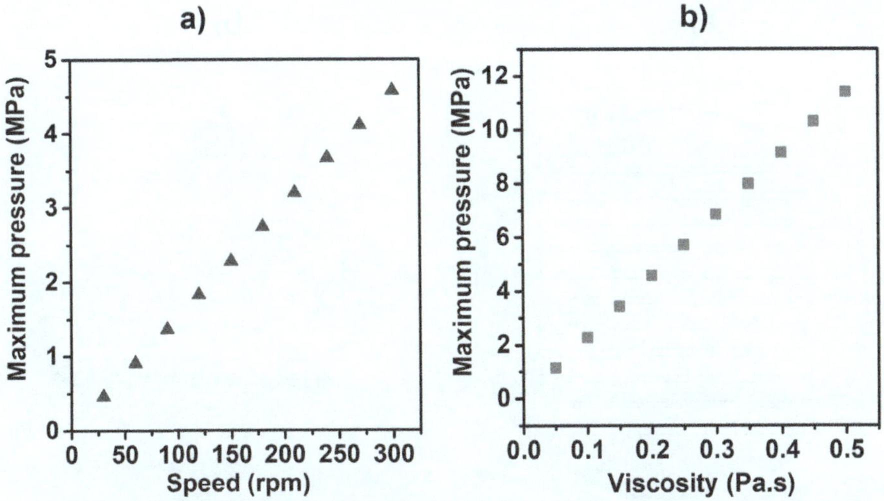

Figure 6.6 Graphs representing the variation of maximum lubricant pressure with (a) lubricant viscosity and (b) rotational speed of the cylinder.

6.5.2 Groove-textured cylinder

To reduce the friction and to increase the tribological performance of a rotating cylinder when it is subjected to certain loads, it is important to provide surface texturing on the surfaces that are subjected to harsh conditions. Here rectangular grooves of constant dimensions and varying depth are constructed, and their tribological performances are investigated and compared with the non-textured ones. Figure 6.7a shows the top view geometry of the cylinder with grooves. Eleven grooves of equal spacing are constructed. The depth extruded here is 100 μm.

Figure 6.7b shows the groove-textured geometry with base velocity direction. Here the groove depth is fixed at 100 μm and the rotational speed of the geometry is 150 rpm. Figure 6.8a depicts the lubricant thickness through the groove. Here the viscosity is kept at 0.1 Pa·s. As similar to non-textured geometry (Figure 6.3a), the film thickness spreads non-uniformly. This is attributed to the viscosity of the lubricant and the cylinder rotation. The magnitude of the film thickness is observed to be close to that of the non-textured cylinder. This is because we have kept the parameters—lubricant viscosity, applied load and cylinder rotation—constant for both geometries. Figure 6.8b represents the pressure distribution of the geometry. The pressure is observed to be maximum in the range of minimum lubricant thickness, depicting the similarity in pressure distribution with the non-textured cylinder. However, the magnitude of the pressure has increased due to the influence of the groove texturing. The lubricant viscosity is kept at 0.1 Pa·s with an applied load of 120 rpm, and the cylinder is rotating at 150 rpm. The groove depth is kept constant at 100 μm.

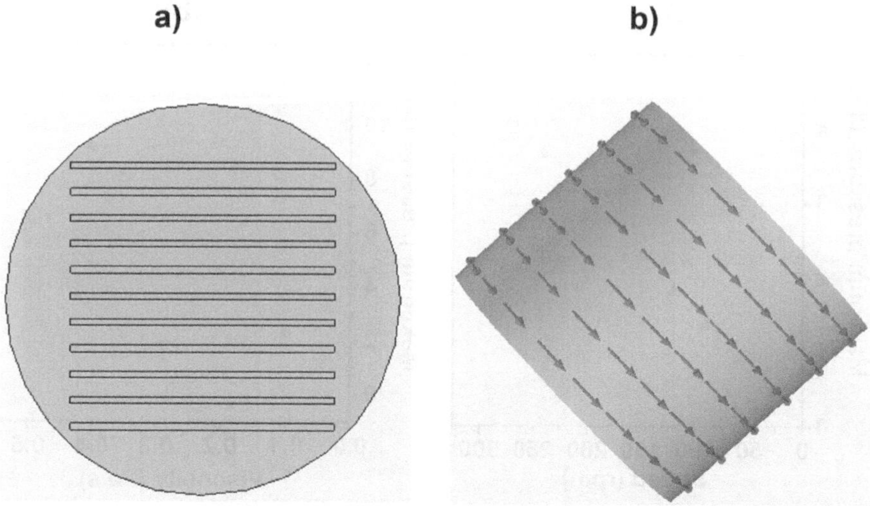

Figure 6.7 (a) Geometry of groove-textured cylinder. (b) Geometry showing the base velocity direction of the groove-textured cylinder with arrows. The depth of the groove is taken as 100 μm.

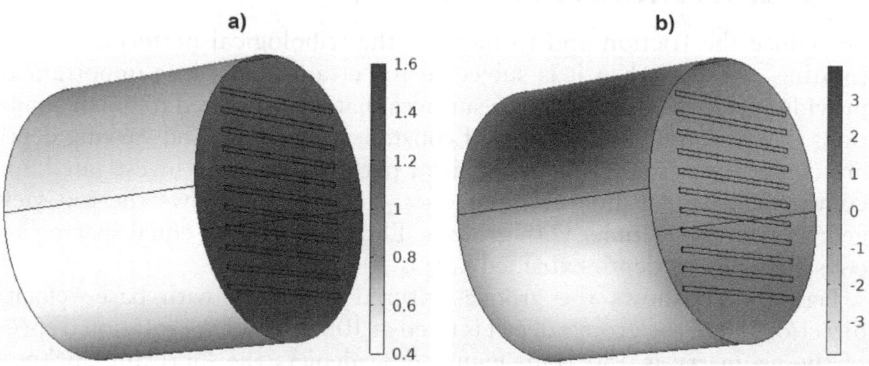

Figure 6.8 Contour representation of (a) the lubricant thickness (mm) and the (b) lubricant pressure distribution (MPa) for the groove-textured cylinder. The depth of the groove is 100 μm.

Similar to non-textured geometry, the normal stress and shear stress distribution are studied for groove-texture and are represented as in Figure 6.9. Surface texturing imposes stress concentrations from straight edges. Thus, the magnitudes of the stresses are estimated to be larger compared to the non-textured cylinder.

Apart from surface texturing, the groove depth also plays a pivotal role in influencing the magnitude of the pressure and various stresses. Figure 6.10 depicts the graphs of variation of maximum normal stress and maximum

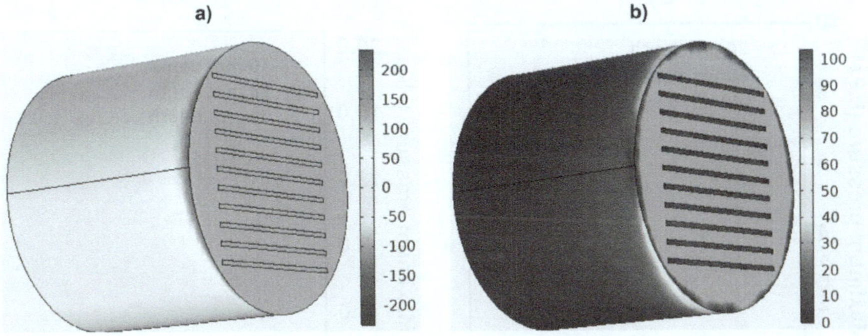

Figure 6.9 Distribution of (a) normal stress and (b) shear stress (in kPa) for a groove-textured cylinder when it is subjected to 120 N applied load with lubricant viscosity of 0.1 Pa·s. The cylinder is rotating at 150 rpm and the depth of the groove is taken as 100 μm.

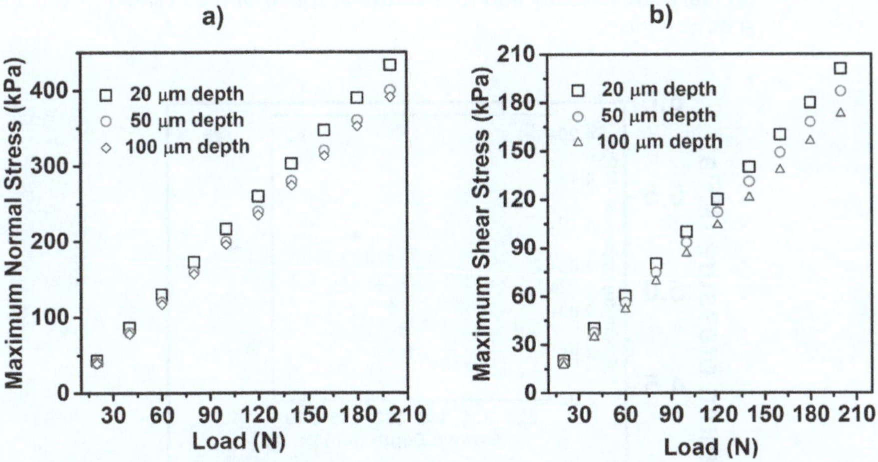

Figure 6.10 Graphs representing the variation of (a) maximum normal stress and (b) maximum shear stress with applied load at various groove depths.

shear stress with applied load at various groove depths. With an increase in the groove depth, the stresses are observed to decrease. Here the viscosity, load and cylinder rotations are kept constant at 0.1 Pa·s, 120 N and 150 rpm, respectively.

Figure 6.11 represents the variation of maximum lubricant pressure with lubricant viscosity and rotational speed of the cylinder at various groove depths. Here the lubricant pressure seems to be constant with an increase in the groove depth. This relation of constant film pressure and decrease in stress with respect to groove depth tempts the researchers to design grooves with the maximum possible depth.

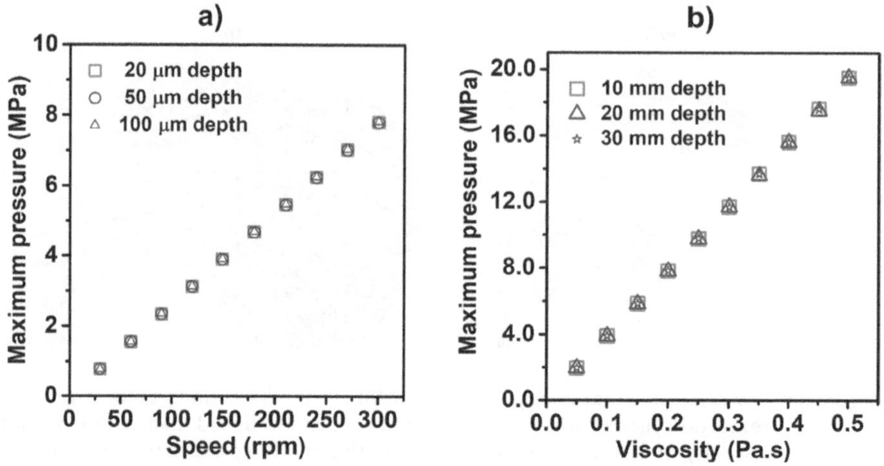

Figure 6.11 Graphs representing the variation of maximum lubricant pressure with (a) lubricant viscosity and (b) rotational speed of the cylinder at various groove depths.

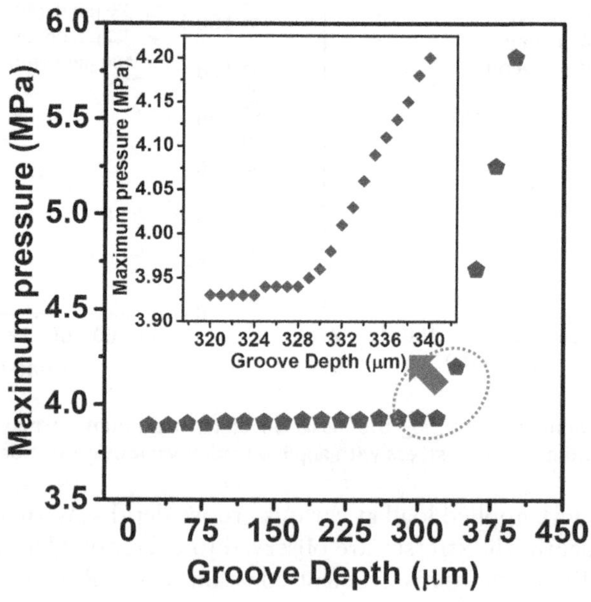

Figure 6.12 Graph representing the variation of maximum lubricant pressure with groove depth.

To investigate whether the lubricant pressure is constant with large groove depths (20 μm to 400 μm), the correlation is plotted as in Figure 6.12. We could observe that the lubricant pressure reduces after a certain groove depth. For an applied load of 120 N to a rotating cylinder (150 rpm) with a lubricant viscosity of 0.1 Pa·s, this groove depth is observed to be 330 μm. Therefore, there is always an optimised groove depth beyond which the lubricant pressure increases rapidly. Obviously, the selection of optimal groove depth depends on various parameters like lubricant viscosity, applied load and rotation of the cylinder.

6.6 CONCLUSION

The tribological performance of the micro-grooved rotating cylinders subjected to varying applied load in the presence of a lubricant of certain viscosity is compared with non-textured rotating cylinders. The lubricant pressure is observed to increase with an increase in lubricant viscosity. Similarly, the lubricant pressure is directly correlated with the rotational speed of the cylinder. When the applied load is increased, the normal stress and the shear stress are also observed to increase. While investigating the relation of these parameters for groove-textured cylinders, it is observed that for a constant groove depth (100 μm), the lubricant pressure and magnitude of various stresses are observed to increase when compared with non-textured systems. However, the stresses are reported to reduce with groove depth. The variation of lubricant pressure is estimated to be constant until 330 μm and then increases. Thus, there is an optimal groove depth which has to be designed for every groove-textured system to ensure better tribological performance as compared to non-textured systems.

REFERENCES

Ali, Saood, Rendi Kurniawan, Park Gun Chul, and Tae Jo Ko. 2023. "Tribological Properties of Hierarchical Micro-Dimples Produced on a Cylindrical Surface by Dual-Frequency Texturing." *Friction* 11 (2). Tsinghua University: 246–58. doi:10.1007/s40544-022-0598-5.

Ammosova, Lena, Kari Mönkkönen, and Mika Suvanto. 2020. "Precise Fabrication of Microtextured Stainless Steel Surfaces Using Metal Injection Moulding." *Precision Engineering* 62 (March). Elsevier Inc.: 89–94. doi:10.1016/j.precisioneng.2019.11.012.

Astakhov, Viktor P. 2013. "Tribology of Cutting Tools." In *Tribology in Manufacturing Technology*, edited by J. Paulo Davim, 1–66. Berlin, Heidelberg: Springer. doi:10.1007/978-3-642-31683-8_1.

Basher, M. K., R. Mishan, S. Biswas, M. Khalid Hossain, M. A. R. Akand, and M. A. Matin. 2019. "Study and Analysis the Cu Nanoparticle Assisted Texturization Forming Low Reflective Silicon Surface for Solar Cell Application." *AIP Advances* 9 (7). American Institute of Physics Inc. doi:10.1063/1.5109003.

Bhaumik, Shubrajit, Viorel Paleu, Dhrubajyoti Chowdhury, Adarsh Batham, Udit Sehgal, Basudev Bhattacharya, Chiradeep Ghosh, and Shubhabrata Datta. 2022. "Tribological Investigation of Textured Surfaces in Starved Lubrication Conditions." *Materials* 15 (23). MDPI. doi:10.3390/ma15238445.

Bompos, D. A., and P. G. Nikolakopoulos. 2016. "Tribological Design of a Multistep Journal Bearing." *Simulation Modelling Practice and Theory* 68 (November). Elsevier B.V.: 18–32. doi:10.1016/j.simpat.2016.07.002.

Ciulli, Enrico. 2019. "Tribology and Industry: From the Origins to 4.0." *Frontiers in Mechanical Engineering* 5. Frontiers Media S.A. doi:10.3389/fmech.2019.00055.

Fan, Hengzhong, Yunfeng Su, Junjie Song, Hongqi Wan, Litian Hu, and Yongsheng Zhang. 2019. "Design of 'Double Layer' Texture to Obtain Superhydrophobic and High Wear-Resistant PTFE Coatings on the Surface of Al2O3/Ni Layered Ceramics." *Tribology International* 136 (August). Elsevier Ltd: 455–61. doi:10.1016/j.triboint.2019.04.004.

Ghani, Jaharah A., Faarih Farhan Mohd Nasir, Haniff Abdul Rahman, Wan Fathul Hakim Wan Zamri, Mohd Shahir Kasim, and Shalina Sheikh Muhamad. 2021. "Computational Fluid Dynamic Analysis on Tribological Performance under Hydrodynamic Lubrication of Dimple Textured Surface Produced Using Turning Process." *Wear* 477 (April). Elsevier B.V.: 203898. doi:10.1016/j.wear.2021.203898.

Haribabu, A., Raviteja Surakasi, P. Thimothy, Mohammad Amir Khan, Nadeem A. Khan, and Sasan Zahmatkesh. 2023. "Study Comparing the Tribological Behavior of Propylene Glycol and Water Dispersed with Graphene Nanopowder." *Scientific Reports* 13 (1). Nature Research. doi:10.1038/s41598-023-29349-7.

Jain, Ankit, and Vivek Bajpai. 2019. "Mechanical Micro-Texturing and Characterization on Ti6Al4V for the Improvement of Surface Properties." *Surface and Coatings Technology* 380 (December). Elsevier B.V. doi:10.1016/j.surfcoat.2019.125087.

Lu, Ping, and Robert J.K. Wood. 2020. "Tribological Performance of Surface Texturing in Mechanical Applications—a Review." *Surface Topography: Metrology and Properties* 8 (4). IOP Publishing. doi:10.1088/2051-672X/abb6d0.

Mao, Bo, Arpith Siddaiah, Yiliang Liao, and Pradeep L. Menezes. 2020. "Laser Surface Texturing and Related Techniques for Enhancing Tribological Performance of Engineering Materials: A Review." *Journal of Manufacturing Processes*. Elsevier Ltd. doi:10.1016/j.jmapro.2020.02.009.

Pietrzyk, Bozena, Katarzyna Porębska, Witold Jakubowski, and Sebastian Miszczak. 2019. "Antibacterial Properties of Zn Doped Hydrophobic SiO2 Coatings Produced by Sol-Gel Method." *Coatings* 9 (6). MDPI AG. doi:10.3390/coatings9060352.

Shen, Gang, Jufan Zhang, Ruslan Melentiev, and Fengzhou Fang. 2021. "Study on Tribological Performance of Groove-Textured Bioimplants." *Journal of the Mechanical Behavior of Biomedical Materials* 119 (July). Elsevier Ltd. doi:10.1016/j.jmbbm.2021.104514.

Teh, Jia Leang, Rashmi Walvekar, Thachnatharen Nagarajan, Zafar Said, Mohammad Khalid, and Nabisab Mujawar Mubarak. 2022. "A Review on the Properties and Tribological Performance of Recent Non-Aqueous Miscible Lubricants." *Journal of Molecular Liquids*. Elsevier B.V. doi:10.1016/j.molliq.2022.120274.

Tung, Simon C., and Michael L. McMillan. 2004. "Automotive Tribology Overview of Current Advances and Challenges for the Future." *Tribology International* 37 (7). Elsevier Ltd: 517–36. doi:10.1016/j.triboint.2004.01.013.

Vishnoi, Mohit, Paras Kumar, and Qasim Murtaza. 2021. "Surface Texturing Techniques to Enhance Tribological Performance: A Review." *Surfaces and Interfaces*. Elsevier B.V. doi:10.1016/j.surfin.2021.101463.

Wu, Wei, Guiming Chen, Boxuan Fan, and Jianyou Liu. 2016. "Effect of Groove Surface Texture on Tribological Characteristics and Energy Consumption under High Temperature Friction." *PLoS ONE* 11 (4). Public Library of Science. doi:10.1371/journal.pone.0152100.

Yin, Yang, Pengfei Shi, Shaohua Zhang, Yuanyuan Jiang, Tao Qing, Yang Wang, Linmao Qian, Jiyang Zhang, and Lei Chen. 2023. "Improving Tribological Performance of Porous Polyimide by Surface Polishing." *ACS Applied Polymer Materials*, June. American Chemical Society (ACS). doi:10.1021/acsapm.3c00795.

Chapter 7

Effect of heat treatment on tribological and mechanical properties of polymeric material

Kamlendra Vikram, Sumit Pramanik, and Shubrajit Bhaumik

7.1 INTRODUCTION

It is well accepted that heat treatment is one of the most efficient methods for altering polymers and increasing their application. The study concluded that heat treatment of polyamide (PA) increased the material's tribological characteristics [1]. In modern times, a great deal of focus has been placed on the process of fabricating a type of absorbing material to find a solution to the issues that are caused by traditional microwave-absorbing materials, such as their inability to be used in environments with high temperatures, lack of mechanical properties, heaviness, and narrow absorption bandwidth [2]. In today's rapidly developing world, composite materials are becoming more essential due to their attractive features. Composite materials made of fibre-reinforced polymer (FRP) have been considered as potential replacements for metal components in various industries over the past few decades, including the aerospace, defence, automotive, electronic, and manufacturing sectors of electronic items. FRP composites provide a high strength-to-weight ratio, better fatigue performance, greater corrosion resistance, and lower density (ρ) than metals [3]. The structural components must be designed around the FRP capacity to operate effectively under various loading circumstances, including torsional, axial, and impact loading. One particular subcategory of polymer matrix composites (PMC) is fibre-reinforced plastic composites. PMC generally consists of two phases: the first one is the matrix phase and the second is the reinforcement phase. Several essential criteria influence the mechanical and tribological properties of FRP composites. The characteristics of these FRP composites may be enhanced by modifying the matrix phase and including appropriate fillers [4, 5]. Surface treatment consists of two stages: removing the damaged fibre layer and increasing the number of surface-active groups. Surfaces may be treated in a variety of ways: via oxidation, chemical scratching, plasma scratching, chemical vapour sedimentation, gamma rays, or ionic induction [6–12]. A chemical treatment is determined mainly by how much oxidizing agent is used, how long it is left to work, how hot it gets, and what kind of

 DOI: 10.1201/9781003713364-7

fibre is being treated. A few procedures may improve fibre–matrix adhesion. The following guidelines may be used to achieve these enhancements [13, 14]:

- By enhancing the wetting properties of fibre surface with the assistance of the matrix resin, we may make the yarn more water-resistant.
- By eliminating contaminating gas molecules physically adsorbed on the fibre surface. This would increase the van der Waals force between the fibre and polymer despite its limited range.
- Incorporating matrix molecules into the molecular network of the polymeric coating on the fibres by physical entanglement or diffusion.
- By enhancing fibre–matrix mechanical interlocking. Surface porosity allows resin molecules to infiltrate and accomplish this purpose.
- The number of chemical reactions between the unreacted species in the matrix resin and the fibre's surface might be increased merely by increasing the number of active sites on the fibre's surface.
- By adding a thin layer as a "coupling agent" that reacts chemically with both the fibre and the matrix, we may achieve this result.

It was determined that heat treatment has a positive influence on the coefficient of friction (COF) and wear rate (WR) characteristics of polytetrafluoroethylene (PTFE), polymethyl methacrylate (PMMA), polyamide (PA), and polyethylene terephthalate (PETP). After 12 hours at 100°C, polymers were cooled using cooled air, H_2O, and a furnace. The test adhesive and abrasive wear resistance, scratch experiments were examined. The results of the experiments show that heat treatment substantially affects the COF and WR of the polymers that were investigated for this study. Regarding abrasive wear, a significant decrease in COF and WR occurred as the cooling rate increased. This was caused by an increase in the polymer plastic deformation, which resulted from the improvement in the cooling rate. As a direct consequence, the water-cooled test specimens experienced the lowest levels of wear and friction. PETP exhibited the lowest wear values of all the adhesives tested, followed by PA, PMMA, and PTFE. PMMA had the highest wear values. Tran et al. demonstrated that carbon fibre (CF) could be modified by boiling it in HNO_3 for 5 hours, which increased the fibre surface O_2 and N_2 content and, in turn, its surface energy [15]. Zhao et al. used hot compression to make CF-reinforced PA6 laminates [16]. The compatibility and interaction with PA6 were increased by adding polyurethane dispersion (PUD) to carbon plain-woven. The flexural and low-cycle flexural fatigue tests revealed that treated CF/PA6 laminates had improved flexural strength and residual flexural strength retention following fatigue tests.

In this chapter, an investigation was conducted into the effect of different types of heat treatment with various volume ratios on the surface properties of polymeric material. Also explored was the impact of a changed polymer on mechanical and tribological characteristics.

7.2 FILLER MATRIX AND POLYMERIC MATERIALS

Approximately 87% of all manufactured polymers are thermoplastics. Based on their typical transition temperature (Tg), they are classified into two categories: amorphous or crystalline [17]. The types of material, application, characteristics, and uses of plastic material are depicted in Table 7.1 [18, 19]. Over 70% of thermoplastics are made from low-cost commodity resins. These materials include different-density polyethylenes (PE), isotactic polypropylene (PP), polystyrene (PS), and polyvinyl chloride (PVC). In terms of performance and price, acrylics, acrylonitrile-butadiene-styrene (ABS) terpolymers, and high-impact polystyrene rank first, second, and third, respectively. Plastics intended for engineering, such as acetals, PA, polycarbonate, polyesters, and polyphenylene oxide, are used in high-performance applications. The technique and production of plastics/polymer materials are illustrated in Figure 7.1, along with some examples of the finished products. Figure 7.2 depicts polymer classification and characteristics.

7.3 EFFECT OF FILLER/REINFORCEMENT

Due to their undesirable geometrical characteristics, surface area, or surface chemical composition, fillers could only moderately raise the polymer's modulus while the strength (tensile, flexural) remained the same or reduced. This was attributable to filler geometry, surface area, and chemical composition. Reinforcing filler is a new word for discontinuous additives whose form, shape, and surface chemistry have been altered to improve polymer mechanical characteristics, particularly strength. To improve compatibility and interfacial adhesion with the chemically different polymer matrix, reinforcing fillers are developed by adjusting the method or the materials used. In addition to enhancing the filler's primary purpose, which in this example is to modify mechanical properties, such adjustments may also introduce or improve new functionalities.

The fillers can be identified using many methods, which are determined by the numerous qualities that fillers present. According to their level of aggregation, all fillers that have been discovered can be placed into one of three distinct categories: gaseous, liquid, or solid [20] (Table 7.2).

Table 7.1 Types of materials, application, and common uses for plastic materials

Material type	*Features*	*Typical applications*	*Tensile strength (MPa)*	*Tensile modulus (GPa)*	*Impact strength (J/m)*	*Density (ρ)*
Acrylics [poly(methyl methacrylate), PMMA]	Excellent light transmission and weather resistance; poor mechanical characteristics	Aircraft enclosures, lenses, drafting equipment, outdoor displays	72.4	3	21	1.19
Fluorocarbons (PTFE or TFE)	Chemically inert in practically all situations, good electrical conductivity, moderate friction; may be utilized to 260°C (500°F); weak cold-flow qualities	Anti-corrosive seals, chemical pipelines and valves, bearings, anti-adhesive coatings, high temp electronics	17.1	36	173	2.2
PA6	High mechanical strength and toughness; low COF; high absorption of water and certain other liquids	Bearings, gears, cams, bushings, handles, and wire/cable jacketing	81.4	2.76	59	1.13
Polycarbonates (PC)	Stable, reduced moisture absorption, clear, high impact resistance and ductility; poor chemical resistance	Safety helmets, lenses, light bulbs, film base	69	2.3	693	1.20
Polyethylene (PE)	Chemically and electrically resistant; tough and low friction; low strength and weak weathering resistance	Flexible bottles, toys, tumblers, battery components, ice trays, film	20-38	0.18-1.6	15-25	0.80-0.96
Polypropylene (PP)	Good in electrical qualities and fatigue strength; chemically inert; cheap; weak UV light resistance	Bottles, film, TV cabinets, luggage	35.8	1.6	43	0.90
Polystyrene (PS)	Good in electrical and optical qualities; strong thermal and dimensional stability; cheap	Battery cases, toys, indoor illumination panels, appliance housings	45.1	3.1	59	1.05

(Continued)

Table 7.1 (Continued) Types of materials, application, and common uses for plastic materials

Material type	*Features*	*Typical applications*	*Tensile strength (MPa)*	*Tensile modulus (GPa)*	*Impact strength (J/m)*	*Density (ρ)*
Vinyl (PVC)	Optimal low-cost general-purpose materials; hard but flexible with plasticizers; typically copolymerized; heat-deformable	Flooring, pipe, wire insulation, garden hose, records	44.4	2.75	181	1.4
Polyester (PET or PETE)	Best in fatigue and tear strength; resistant to dampness, acids, greases, oils, and solvents	Tapes, clothes, tire cords, bottles	159	9.96	101	1.56

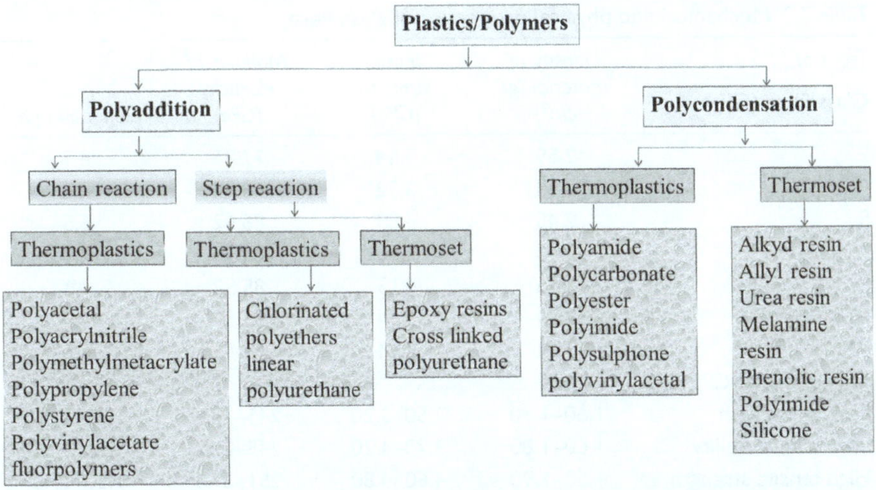

Figure 7.1 Techniques for the production of plastics, along with some examples.

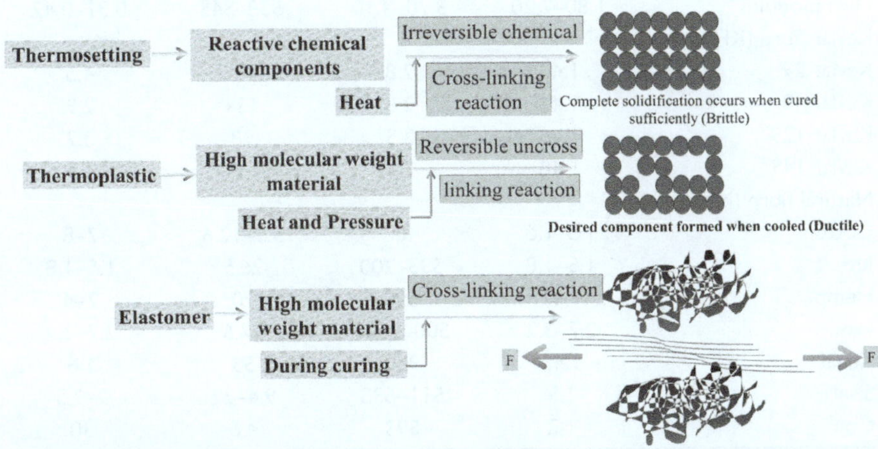

Figure 7.2 Polymer classification and their characteristics.

7.4 SURFACE TREATMENT TECHNIQUE ON POLYMERIC MATERIALS

7.4.1 Chemical treatment

Wet chemistry coatings are a major chemical technique for surface modification. The coating application/consolidation temperature is another key applicability requirement due to the small temperature range to which plastics may be subjected without physical deformation or chemical deterioration [21]. As the moisture content rises, the dissolvable character of the

Table 7.2 Mechanical and physical properties of glass fibre

Types of filler Glass fibre (GF)	*Density of material (g/cm³)*	*Tensile strength (GPa)*	*Modulus of elasticity (GPa)*	*Elongation (%)*
E	2.59	3.44	72.31	4.7
C	2.53	3.32	68.94	4.6
S	2.45	4.88	86.92	5.5
A	2.43	3.30	68.99	4.9
R	2.56	4.13	85.58	49
EC	2.74	3.44	85.57	4.7
AR	2.72	3.24	73.11	4.5
Carbon fibre (CF)				
Low in strength	1.60–1.90	2.50–3.20	245–275	1.20–1.50
Strong in its ability	1.60–1.80	3.70–4.20	245–275	1.20–1.30
High tensile strength	1.50–1.90	4.60–4.80	251–272	1.20–1.45
Intermediate modulus	1.80–1.90	5.80–6.10	305	1.21–1.42
Superior modulus	1.60–1.80	2.40–3	382–405	0.61–0.73
High modulus	1.80–2.20	3.70–4.10	538–545	0.31–0.42
Kevlar fibre (KF)				
Kevlar 29	1.43	2.8	71	4.5
Kevlar 49	1.46	2.8	134	2.9
Kevlar 129	1.46	3.5	98	3.2
Kevlar 149	1.48	2.4	144	1.6
Natural fibre (NF)				
Cotton	1.5–1.6	400	5.5–12.6	7–8
Jute	1.5–1.8	393–700	26.5	1.5–1.8
Hemp	1.47	690	70	2–4
Flax	2.7–3.2	500–1500	27.6	2.7–3.2
Kenaf	1.45	930	53	1.6
Sisal	1.5	511–635	9.4–22	2–2.5
Coir	1.2	593	4.6	30

compounds of acids and base solvents on the surface of polymeric materials increases. There has also been work done with acid treatments in order to improve the interfacial characteristics of CF-reinforced polymers thermoplastic (CFRTP) composites. Khan et al. revealed how high-density polyethylene (HDPE) and CF react without a coupling agent [22]. Based on their research, they determined that an increase from 10 to 20 CF layers in composites significantly enhanced mechanical properties. Shengbo et al. observed that HNO_3 chemical treatment of CF is more useful with HDPE than untreated CF [23]. They reported that the compatibility of the treated CF with the PE matrix was improved due to an interaction between the hydrophilic carboxylic group of benzoic acid and a hydroxyl group of the CF [23]. Chunzheng et al. demonstrated that acid treatments strengthened the ultrahigh-molecular-weight polyethylene (UHMWPE)–CF interface [24]. Zhang et al. conducted an experiment in which CF was oxidized using nitric acid and subsequently embedded in polyoxymethylene composites (POM/CF) [25]. Finally, they observed that nitric acid-treated fibre had a disorganized, rough surface and reactive functional groups. Nitric acid treatment improved the fibre surface roughness, increasing POM/CF composites' flexural strength and modulus [25]. Shengbo et al. employed the electrochemical approach to change the surface of the CF [23]. Nitric acid was used to treat packs of CF that had been linked to positive electrodes (HNO_3). It was done to change the CF surface. It suggests functionalizing the CF surface increased interfacial strength [23]. Li et al. found that treating carbon fibre-reinforced ABS composites with HNO_3 enhanced interfacial adhesion [26]. Our other studies additionally demonstrated that the tribological and mechanical characteristics of the glass fibre reinforcement, the interfacial bonding between the PA6 matrix and the GF, and the graphite reinforcement had improved outcomes after being chemically treated with nitride acid (HNO_3) [27, 28]. Li et al. conducted an empirical study on two different approaches with the goal of bettering the interfacial interaction between CF and PS [29]. PS was grafted with maleic anhydride to make it more compatible with the CF, and nitric acid was employed to provide the CF surface new functions. According to the authors, enhancing the surface O_2 and N_2 increased the fibres' total surface energy, which produced good interfacial contact between carbon fibre and polystyrene matrix. This finding was consistent with what was anticipated to occur when the surface of carbon fibre was treated [29]. Using the oxidation technique, Yan et al. changed the surface of CF, and then coated it with PA12 to protect it. Both physical and chemical approaches (coating and oxidation) were employed in this process [30].

7.4.2 Corona discharge

Figure 7.3 shows how plastic film is corona-treated by passing it between an electrode and a dielectric-coated earthed roller. The electrode and roller have a 23 mm air gap. When a film passes over a corona discharge, the surface molecules oxidize, increasing the polymer's surface tension. The increase in OH and COOH groups on the polymer surface, which is connected to surface polarity, increases surface stickiness and wettability.

7.4.3 Flame treatment

Flame treatment is used to improve the adhesion properties of polymers before they are coated or adhered to. Flame-treated surfaces are superior

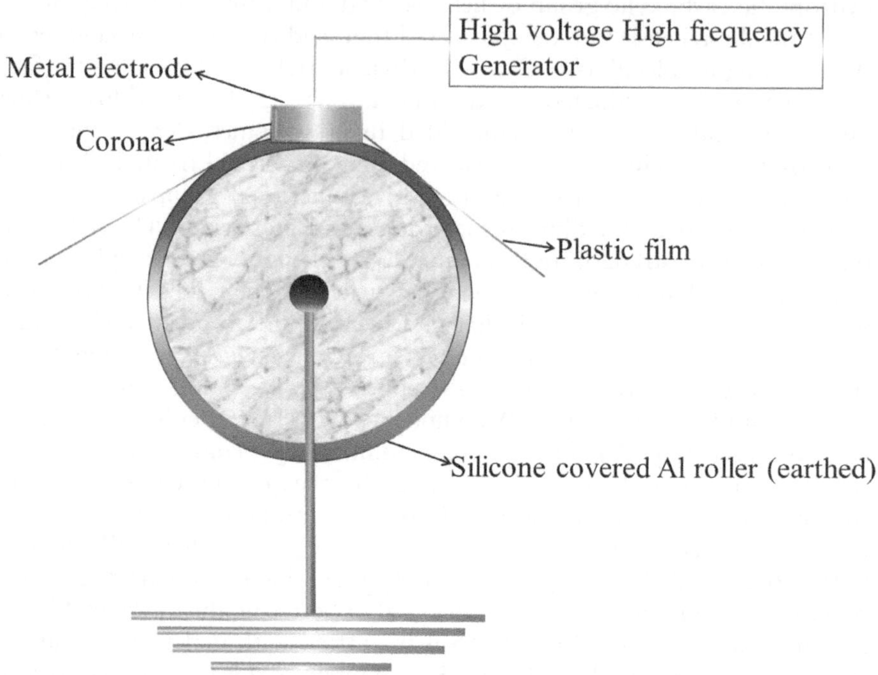

Figure 7.3 Corona treatment process.

in terms of bond strength and durability. In fact, flame treatment is more stable than corona treatment. As a consequence of flame treatment's cheap cost, total production costs are reduced. Flame treatment is used in many sectors, including the automotive industry to improve paint adherence on polypropylene mouldings (such as vehicle bumpers). Figure 7.4 depicts the primary equipment components. The quantity of oxygen (O_2) added to polymer surfaces by flame treatment is linked to the number of passes and length of exposure.

7.4.4 Self-assembled monolayer (SAM)

Self-assembly is the process by which a disordered system of components forms and organizes a structure or pattern as a consequence of specific, local interactions among the components themselves, without any direction from the outside. This process can take place even in the absence of any outside influence. This procedure is carried out in the absence of any advice from outside sources. In the context of surfaces, this suggests that some molecules are able to interact with a surface and produce a self-assembled monolayer on the surface, which is also referred to as SAM. The chemical structure of the molecules that comprise the SAM layer allows for the

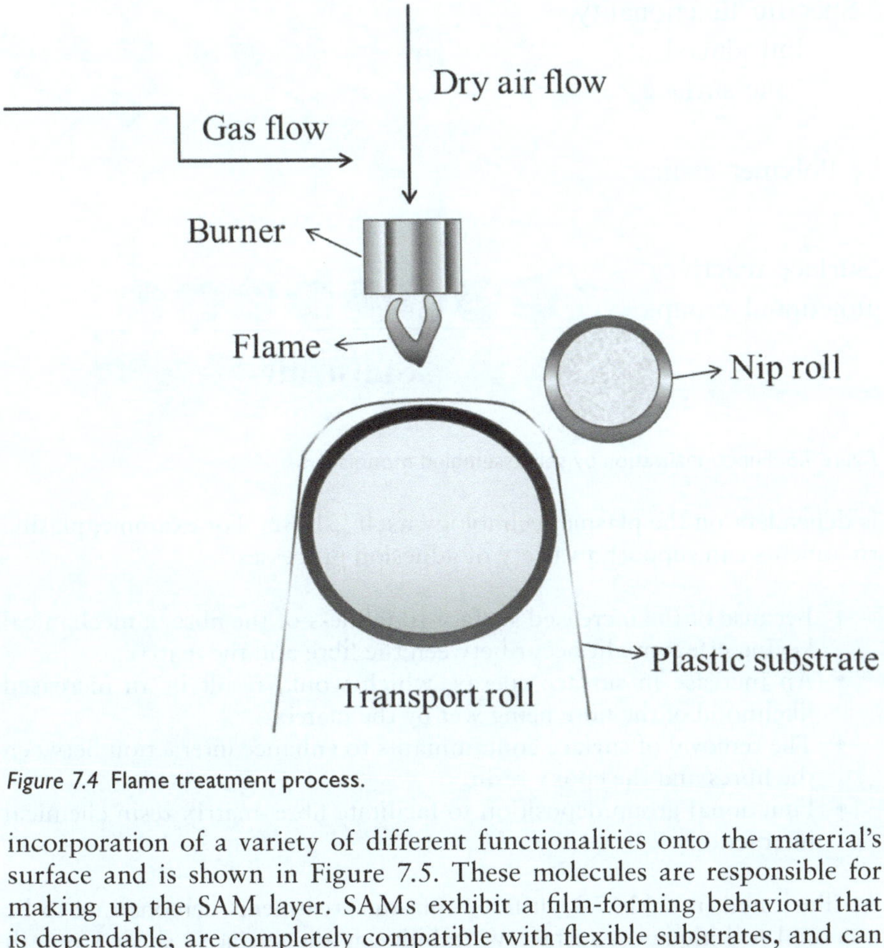

Figure 7.4 Flame treatment process.

incorporation of a variety of different functionalities onto the material's surface and is shown in Figure 7.5. These molecules are responsible for making up the SAM layer. SAMs exhibit a film-forming behaviour that is dependable, are completely compatible with flexible substrates, and can be processed at a cheap cost. SAMs provide a significant benefit to industries including tissue engineering, biomechanics, synthetic biology, and biomedical devices, where the interplay between biological surroundings and regularly structured surfaces plays a critical role. These fields include tissue engineering, biomechanics, and synthetic biology.

7.4.5 Plasma treatment

Plasma treatment is one way that can be used to raise the surface energy of polymeric materials, particularly those with low energies (Figure 7.6). The treatment modifies the surface's thermal appearance, which in turn enhances the overall adhesion that may be achieved by wetting liquids. Better adhesive bonding and more manageable viscosity effects in composites can be achieved with plasma treatment, and this treatment connection

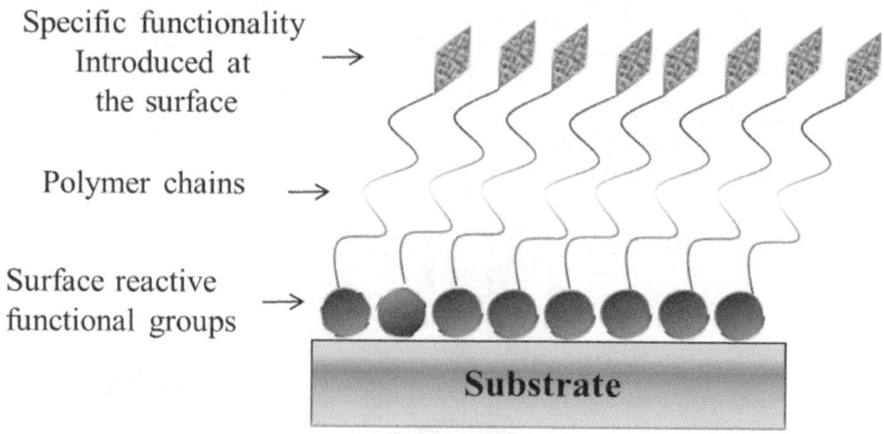

Figure 7.5 Functionalization by self-assembled monolayers.

is dependent on the plasma technology itself [31, 32]. For example, plasma treatments can support a variety of adhesion processes:

- Because of the increased surface roughness of the fibre, a mechanical keying effect might occur between the fibre and the matrix.
- An increase in surface energy, which would result in an increased likelihood of the fibre being wet by the matrix.
- The removal of surface contaminants to enhance interaction between the fibres and the epoxy resin.
- Functional group deposition to facilitate fibre–matrix resin chemical interactions.

As illustrated in Table 7.3, different plasma treatments demonstrate to be a crucial technique on surface wettability as the surface's shear strength is higher than the surface treated conventionally [32, 33]. The strategy of utilizing various approaches in adaptive circumstances might be differentiated between standard pressure and moderate pressure.

7.4.6 Nanoparticale grafting

A nanostructured surface may increase cell adherence by providing focal locations for filopodia to attach. As a result, the surface will replicate the critical aspects of the morphology of an extracellular matrix, which will help the cells to adhere to the surface. The grafting of nanoparticles or biomolecules to a polymer substrate is a multistep procedure that involves modifying the surface's physico-chemical properties and shape to ensure good contacts and bonding sites [34]. There are variable degrees of activation on the surface. For example, PET foils grafted containing noble

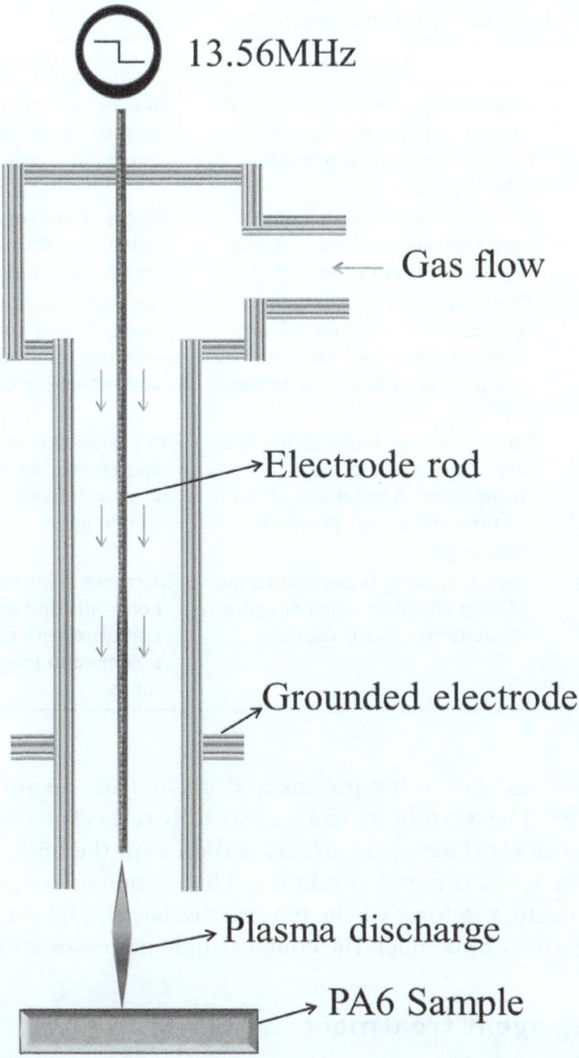

Figure 7.6 Plasma heat treatment process.

metal nanoparticles (Pt and Pd) required plasma treatment to establish enough bonding sites. The biphenyl-4,4′-dithiol (BPD) binding agent is used to attach platinum and palladium nanoparticles to the polymer substrate. The BPD occupied the active sites made by plasma therapy. After the procedure, mammalian cells were transplanted [35]. Plasma pretreatment dissolves macromolecular bonds on the modified polymer's surface, generating highly reactive free radicals. These radical active sites are quickly attacked by oxygen molecules in the atmosphere, resulting in the formation

Table 7.3 Methods of surface plasma treatments

Method	*Description*	*Treatment*
Microwave plasma	There is a significant amount of ion and radical absorption. Ensures the highest possible etching rates.	Flexible operating conditions lead to increased surface energy and pollution.
Induction plasma	Restored by the movement of gas. The temperature that was measured was rather high.	Effects of etching, including surface oxidation and increased roughness, improve wettability.
Radiofrequency plasma	The flow of gas that is being exerted by the RF field. The ratio of the power absorbed by the plasma to the total incident power.	Wettability enhancement is a reasonably simple and cost-effective method for advancing and retreating contact angles.
Active gas plasma	Plasma torches fed with Te~Tgas are responsible for its production. A extended amount of time and a high-pressure discharge.	To change the surface physically and chemically, dissociate gas in a vacuum using an electric field of energy.
Oxygen plasma	The O_2 is being brought into the plasma chamber, enhancing the capacity to absorb water.	Increases in surface oxygen concentration are shown to correlate with higher discharge powers and longer treatment times.

of various oxygen-containing groups and an increase in surface polarity and wettability. The wettability changes steadily reverse as the samples age, and the polymer's surface polar groups realign with the bulk. This restores the wettability to its original condition. The chemical composition of the polymer substrate, the force of the plasma discharge, and duration of heat treatment all affect how much the contact angle increases with time [36].

7.4.7 Sizing agent treatment

In order to alter the composition and characteristics of the composite interface layer, a layer of polymer sizing agent is applied to the fibre surface. This process not only makes the interface more compatible but also does not harm the fibre. This method compares favourably to the described methods for treating the fibre surface. At the same time, it has better interface bonding capability, and it does so without sacrificing any of its other in-plane mechanical features [37].

Treating fibre surfaces may increase reinforcement–matrix adhesion [38]. Untreated reinforcements result in weaker interlaminar shear strength in composites [39]. These conditions cause poor matrix–fibre adhesion and bonding, which improves the reinforcement's wettability [40]. Increased roughness leads to a more significant fibre or reinforcement surface area, which increases matrix and reinforcement interaction [41]. The surface treatment may affect the reinforcement's chemical structure, which enhances the matrix–reinforcement chemical link. The treatment efficiency in improving reinforcement surface quality varies with oxidative medium concentration, temperature, treatment duration, and reinforcement [42]. The majority of the time, oxidation is carried out utilizing gases, such as air, oxygen (O_2), ozone (O_3), or liquids like nitric acid (HNO_3). Treatments for oxidation include gas-phase, catalytic, and liquid-phase oxidation (chemically or electrochemically). The components of plasma include neutral atoms or molecules, negatively charged electrons, and positively charged ions [43]. Plasma surface treatment of reinforcements increases surface reinforcement–matrix bonding strength without affecting mechanical qualities [14]. Plasma treatments may prompt different adhesion mechanisms:

- By increasing the surface energy, which enhances the matrix wetting of the reinforcement.
- By increasing the surface roughness of the fibre, which leads to a mechanical locking of the matrix and the fibres.
- By removing any impurities from the reinforcement's surface.
- By adding functional groups to provide optimal chemical interactions between reinforcement and matrix [44, 45].

Fibre strengthening in polymer composites enhances with radiation. This is owing to the fibre's rougher surface and improved adherence to the matrix [6].

The material's characteristics improve if radiation is applied to fibre reinforcement in polymer composites. This is due to the roughening of the fibre surface and improved adherence between the fibre and matrix. Fibre orientation is one element considered in the tribology of fibre-reinforced composites. Several studies use various orientations, including normal, parallel, antiparallel, and random. Regular orientation has the lowest WR, whereas random orientation has the greatest [46, 47]. Table 7.4 illustrates how heating materials influence the tribological characteristics of fillers. The volume proportion of composite fibres influences the material's load-bearing capability. The fibre's load-bearing capability increases with its volume fraction. The capacity of fibres to sustain a weight improves up to a specific volume proportion; beyond that, it decreases. This observation stands in contrast to widespread assumptions. This is because fibre and resin lose their connection over time [48, 49]. Dai et al. have attempted to examine how heat treatment impacts the surface characteristics of CF and the interfacial adhesion between fibres and epoxy. The interfacial connection of T300B and T700SC fibres was reduced by the presence of the sizing agent on their surfaces. This

Table 7.4 Tribological performance of composite materials made of polymeric materials

Polymer	*Fibre*	*Filler/treatment*	*Developed method*	*COF*	*WR*	*Mechanism*	*Reference*
Polyester	GF		Moulding	↓	↓	Plastic deformation	[53]
	Al	SiC	Powder metallurgy		↓		[54]
PEI	Glass fibre	Graphite, MoS_2, PTFE			↓	Adhesion	[55]
Epoxy	CF		Resin transfer mouldings system	↓	↓	Worn surfaces	[56]
Phenolic	Graphite (Grt)		Compression moulding machine		↑↓		[57]
Epoxy	CF		Hand lay-up system				[58]
Polyester	GF		Hand lay-up system	↓	↓		[59]
Epoxy	Rubber dust		Moulding		↓	Abrasive	[60]
Epoxy	GF	SiO_2	Cylindrical mould	↓	↓	Adhesion	[61]
	Al	Red mud	Powder metallurgy	↓	↓	Adhesion	[62]
Epoxy	GF	SiC	Hand lay-up	↓	↓	Debonding, fibre breakage	[63]
Epoxy	GF	Graphite and SiC	Hand lay-up	↓	↓		[64]
Epoxy	Lantana camara		Moulding	↓	↓	Abrasion	[65]
Epoxy	Kenaf		Moulding	↑	↓	Adhesion	[66]
Epoxy	Pin wood dust				↓		[67]
Polyester	Betelnut fruit fibre		Hand lay-up	↑	↓	Crack, debonding	[68]
Epoxy	Pinewood dust		Hand lay-up		↓	Cutting	[69]
Polyester	Linen, jute		Moulding	↑	↓	Fibre debonding	[70]
Polyester	Coir	Alkaline	Hand lay-up	↑	↓	Adhesion	[71]

(*Continued*)

Table 7.4 (Continued) Tribological performance of composite materials made of polymeric materials

Polymer	*Fibre*	*Filler/treatment*	*Developed method*	*COF*	*WR*	*Mechanism*	*Reference*
Epoxy	Borassus fruit fibre	Alkaline	Cylindrical mould	↓	↓	Micro-crack	[72]
Polyester	Oil palm	Alkaline	Cylindrical mould	↓	↓	Debonding	[73]
Epoxy	Bamboo		Hand lay-up	↓	↓		[74]
Epoxy	Rick husk	Banzolylation	Cylindrical mould		↓		[75]
Polyester	Bamboo powder		Cylindrical mold		↓		[76]
Epoxy	Coir	Banzolylation	Cylindrical mould		↓		[77]
Polurethane	Kenaf	Alkaline	Hand lay-up	↓	↓		[78]
Epoxy	Cellulose fibre	Nano SiC, Nano clay			↓		[79]

is against the broad rules that have been set. Desizing decreased the carbon fibre surface acid parameter, which increased the binding strength between the fibre and epoxy at the point of contact [50]. In a recent study, Zhang and colleagues oxidized polyacrylonitrile (PAN)-based CF with strong HNO_3 at 90°C for 1.5 hours. Their experiments found that treating the fibre's surface made it rougher and that the concentration of O_2 increased significantly, both of which contributed to better adhesion between the fibre and the PI matrix. To do this, the PAN-based CF was heated at 90°C for 1.5 hours [43]. Tran et al. altered CF by boiling them in HNO_3 for 5 hours. According to their findings, the more severe the oxidation, the higher the surface oxygen and nitrogen concentration, and hence the higher the surface energy of the fibres. However, after 9, 12, 25, and 50 hours of refluxing CF in 68% fuming nitric acid, the fibres' exterior surface area increased by a factor of 3.7 due to the formation of edges or active sites [15]. In another study, CF was refluxed in 60% nitric acid for 40, 60, and 80 minutes at a temperature of 100°C. After 60 minutes of treatment, the BET surface area of treated CF is ten times that of untreated CF, as determined by the researchers using the Brunauer–Emmett–Teller technique [51]. Tiwari et al. used a 110°C nitric acid (HNO_3, 65–68%) boiling process to oxidize CF. The acid treatment lasted 15 to 180 minutes, and fibre surface roughness increased with time. Spectra between 950 cm^{-1}, 1200 cm^{-1}, 1650 cm^{-1}, and 1710 cm^{-1} for treated fibres revealed ether, carboxyl, and carbonyl groups [52]. Table 7.5 depicts the thermoplastic polymer, CF modification, and mechanical characteristics of various CFRTP composites. Jang and Yang studied oxygen plasma therapy's influence on CF surface morphology over 1 to 5 minutes. The BET surface area on the CF surface increased due to the micro-pitting, with the maximum value reaching after 3 minutes of treatment. The treatment led to an average 20–50% improvement in all mechanical properties. FTIR-ATR and XPS tests show that the cold, remote nitrogen-oxygen plasma (CRNOP) treatment added polar oxygenated functional groups to the CF surface [51].

7.5 CONCLUDING REMARKS

According to the findings of a study that was conducted, there is currently a growing interest in the process of heat-treating polymer-based composites for technological applications that require low friction as well as wear. This interest can be attributed to the fact that these applications are becoming increasingly common. The evaluation demonstrates that the material's mechanical characteristics, such as fatigue resistance, fracture toughness, tensile strength, wear resistance, and friction coefficient, have been improved. Additionally, the friction coefficient has decreased. The results of the study included:

Table 7.5 Thermoplastic polymer, carbon fibre modification, and mechanical characteristics of various carbon fibre reinforced thermoplastic composites

Composite material	*CF modification method*	*Mechanical properties*	*Reference*
PEI/CCF	—	Significant improvements were observed in Young's modulus, toughness, and percent strain perpendicular to the loading direction.	[80]
PP/RCF	Plasma treatments after coupling agents	ILSS might increase by as much as 47.8%.	[81]
CBT/carbon fibre	Treatments with microwave plasma	Increase of about 436.3% in tensile strength	[82]
PP/short carbon fibre	After sizing the materials, plasma treatment will be performed. All of the samples were treated with a coupling agent.	An up to 47.8% increase in ILS value.	[83]
PP/LCF	Coupling agent	The composite mechanical strength was greatly enhanced as a consequence.	[84]
Carbon fibre/ PVDF	Coupling agent	There have been considerable gains in the flexural strength and modulus of 47% and 74%, respectively.	[85]
Carbon fibre/ PA6	Coupling agent and sizing treatment	In contrast to the tensile strength of other composites that included short carbon fibre, the tensile strength of composites that contained long carbon fibre was 24% more.	[86]
Carbon fibre/PC	Sizing treatment	Significant improvements were seen in both the modulus and tensile strength, with the latter increasing by 540% and the former by 150%	[87]
PPEK/carbon fibre	Sizing treatment	When compared to the unsized CF, which had an ILSS value of roughly 39.50 MPa, the sized CF's value is higher at about 51.50 MPa.	[88]

(*Continued*)

Table 7.5 (Continued) Thermoplastic polymer, carbon fibre modification, and mechanical characteristics of various carbon fibre reinforced thermoplastic composites

Composite material	*CF modification method*	*Mechanical properties*	*Reference*
PPEK/carbon fibre	Sizing treatment	The residual radial compressive stress at the fibre/matrix interface was responsible for about 80% of the ILSS in the PPEK/CF composite system.	[89]
PC/chopped carbon fibre	Coating of the material, followed by treatment with a coupling agent	By adding more and more CF, it was possible to see a distinct shift from ductile to brittle behaviour in the composites, as well as an increase in their strength and modulus.	[90]
HDPE/carbon fibre	Without using any coupling agent	The flexural characteristics of the composites have seen a considerable improvement as the amount of CF used in their construction increased.	[22]
POM/carbon fibre	Oxidation treatments	Both the flexural strength and the modulus have seen significant improvements.	[25]
PA12/carbon fibre	Oxidation method followed by coating with a layer of PA12	Both the flexural strength and the modulus saw significant increases, with 114% and 243.4% increases, respectively.	[30]
UHMWPE/carbon fibre	Acid treatment	The ILSS was found to have increased by 70%.	[91]
PA6/ABS/short carbon fibre	Acid treatment	Both the tensile strength and the tensile modulus increased to a significant extent. On the other hand, when PA6 was mixed together with ABS, these qualities have seen a significant improvement.	[26]
PI/carbon fibre	Ozone modification and air-oxidation modification	The friction and wear characteristics of the composite have been improved.	[92]
PP/RCF	Different plasma powers treatment	The values of tensile and flexural strength of composites improved significantly by 17% and 11%, respectively, when the load was raised to 100 W.	[93]

(*Continued*)

Table 7.5 (Continued) Thermoplastic polymer, carbon fibre modification, and mechanical characteristics of various carbon fibre reinforced thermoplastic composites

Composite material	*CF modification method*	*Mechanical properties*	*Reference*
PP/carbon fibre	Irradiated PP as compatibilizer agent	There was a 30% increase in tensile strength.	[94]
PP/short carbon fibre	Material coating	There was a 43% increase in the combined flexural, tensile, and impact strength.	[95]
PA6/carbon fibre	Material coating	The ILSS had a 40.2% growth.	[96]
PES/carbon fibre	Material coating	There was a maximum increase of 12.1% in tensile strength, 31.7% in Young's modulus, 12.4% in flexural strength, and 17.3% in flexural modulus.	[97]
PP/carbon fibre	Material coating	The ILSS observed 300% growth.	[98]
PEEK/carbon fibre	Material coating	The strength has increased by 115.4%, while the resistance to impact has increased by 27%.	[99]
PTFE/CF and PPS/carbon fibre	Sizing materials	The material hardness, wear resistance, elongation at break, strength, and modulus were significantly improved.	[100]
PTT/carbon fibre	Sizing materials	There was a discernible improvement in tensile strength of up to 120% and flexural strength of up to 30%.	[101]
PP/short carbon fibre	Sizing materials	Even while fibre length, tensile strength, tensile modulus, and impact strength all decreased as the quantity of SCF increased, tensile strength, tensile modulus, and impact strength all increased. This was the case even though the short carbon fibre content increased.	[102]
PEEK/carbon fibre	Coupling agent	Both the tensile strength and the modulus saw significant increases, with the former rising by 455% and the latter by 168%.	[103]
Carbon fibre/PA6	Coupling agent	The ultimate tensile strength, elastic modulus, and elongation at break values were all much higher than average.	[104]

- The COF varied more when the heat treatment temperature was increased.
- As cooling rates were reduced, polymers performed better in terms of wear. In order to maximize its protective impact against wear, heat treatment should be applied at increasingly high temperatures.
- The altered crystallization of polymers after heat treatment is responsible for the resulting changes in friction and wear characteristics.
- In general, the heat treatment of the reinforcement in many thermoplastic composites resulted in a significant enhancement in the characteristics of the finished product in contrast to the attributes of the pure thermoplastic.
- To attain the required characteristics, most notably the mechanical and tribological properties, as well as to understand the links between the modification procedures and the mechanical properties of the heat-treated polymeric material, it is vital to consider all of these elements.

AUTHOR CONTRIBUTIONS

The manuscript was written by K.V. and reviewed by S.P. and S.B. All authors contributed to the interpretation of the data and revision of the final manuscript.

INFORMED CONSENT STATEMENT

Not applicable.

DATA AVAILABILITY STATEMENT

Not applicable.

CONFLICTS OF INTEREST

The authors declare that there is no conflict of interest.

REFERENCES

1. Ayman A, Zeidan E, Hamed A, Ali W. Effect of heat treatment on the abrasion resistance of thermoplastic polymers. *Journal of the Egyptian Society of Tribology*. 2010;7(4):52–64.

2. Wang H, Zhu D, Zhou W, Luo F. Influence of heat treatment on electromagnetic properties of polyimide/carbon black composites. *Polymers for Advanced Technologies*. 2014;25(12):1616–21.
3. Jain R, Lee L. *Fiber Reinforced Polymer (FRP) Composites for Infrastructure Applications: Focusing on Innovation, Technology Implementation and Sustainability*. Springer; 2012.
4. Rowell RM. *A New Generation of Composite Materials from Agro-based Fiber. Polymers and other Advanced Materials: Emerging Technologies and Business Opportunities*. Springer; 1995. p. 659–65.
5. Davim JP, Cardoso R. Effect of the reinforcement (carbon or glass fibres) on friction and wear behaviour of the PEEK against steel surface at long dry sliding. *Wear*. 2009;266(7–8):795–9.
6. Tiwari S, Sharma M, Panier S, Mutel B, Mitschang P, Bijwe J. Influence of cold remote nitrogen oxygen plasma treatment on carbon fabric and its composites with specialty polymers. *Journal of Materials Science*. 2011;46:964–74.
7. Baun WL. Anodization of evaporated aluminium on Ti-6 wt% Al-4 wt% V. *Journal Materials Science*. 1980;15(11):2749–2753.
8. Wilson W, Haggmark L, Biersack J. Calculations of nuclear stopping, ranges, and straggling in the low-energy region. *Physical Review B*. 1977;15(5):2458.
9. Wan Y, Wang Y, Huang Y, Luo H, Chen G, Yuan C. Effect of surface treatment of carbon fibers with gamma-ray radiation on mechanical performance of their composites. *Journal of Materials Science*. 2005;40:3355–9.
10. Li J, Huang Y, Xu Z, Wang Z. High-energy radiation technique treat on the surface of carbon fiber. *Materials Chemistry Physics*. 2005;94(2–3):315–21.
11. Tang LG, Kardos JL. A review of methods for improving the interfacial adhesion between carbon fiber and polymer matrix. *Polymer Composites*. 1997;18(1):100–13.
12. Vikram K, Pramanik S, Paleu V, Bhaumik S. Surface modification of magnesium alloy employing external coating for biomedical applications. In Kumar D, Ansari N. *Magnesium Alloys for Biomedical Applications*. CRC Press; 2025. p. 109–23.
13. Jang BZ. Control of interfacial adhesion in continuous carbon and Kevlar fiber reinforced polymer composites. *Composites Science Technology*. 1992;44(4):333–49.
14. Tiwari S, Bijwe J. Surface treatment of carbon fibers-a review. *Procedia Technology*. 2014;14:505–12.
15. Tran MQ, Ho KK, Kalinka G, Shaffer MS, Bismarck A. Carbon fibre reinforced poly (vinylidene fluoride): Impact of matrix modification on fibre/polymer adhesion. *Composites Science Technology*. 2008;68(7–8):1766–76.
16. Zhao D, Hamada H, Yang Y. Influence of polyurethane dispersion as surface treatment on mechanical, thermal and dynamic mechanical properties of laminated woven carbon-fiber-reinforced polyamide 6 composites. *Composites Part B: Engineering*. 2019;160:535–45.
17. Agarwal S, Greiner A, Wendorff JH. Functional materials by electrospinning of polymers. *Progress in Polymer Science*. 2013;38(6):963–91.
18. Maccani R. Characteristics crucial to the application of engineering plastics. *Engineering Plastics*. 1988;2:69.

19. Berins M. *Plastics Engineering Handbook of the Society of the Plastics Industry*. Springer Science & Business Media; 1991.
20. KIRK OTHMER O. *Encyclopedia of Chemical Technology*. WATCHER; 2021.
21. Munawar MA, Khan SM, Gull N, Shafiq M, Islam A, Zia S et al. Fabrication and characterization of novel zirconia filled glass fiber reinforced polyester hybrid composites. *Journal of Applied Polymer Science*. 2016;133(27).
22. Khan SM, Gull N, Munawar MA, Islam A, Zia S, Shafiq M et al. 2D carbon fiber reinforced high density polyethylene multi-layered laminated composite panels: Structural, mechanical, thermal, and morphological profile. *Journal of Materials Science Technology*. 2016;32(10):1077–82.
23. Shengbo Z, Qingling D. The surface modification of carbon fiber for thermoplastic HDPE composites. *Surface Interface Analysis*. 2019;51(2):184–9.
24. Chunzheng P. Improved interfacial properties of carbon fiber/UHMWPE composites through surface coating on carbon fiber surface. *Surface Interface Analysis*. 2018;50(5):558–63.
25. Zhang Y, Zhu S, Liu Y, Yang B, Wang X. The mechanical and tribological properties of nitric acid-treated carbon fiber-reinforced polyoxymethylene composites. *Journal of Applied Polymer Science*. 2015;132(25):41812.
26. Li J, Cai C. The carbon fiber surface treatment and addition of PA6 on tensile properties of ABS composites. *Current Applied Physics*. 2011;11(1):50–4.
27. Vikram K, Bhaumik S, Pramanik S. Effect of graphite on tribological and mechanical properties of PA6/5GF composites. *Journal of Thermal Analysis Calorimetry*. 2023;148(9):3341–55.
28. Vikram K, Pramanik S, Bhaumik S. Effect of hexagonal boron nitride on structural, mechanical, and tribological behavior of polyamide 6/glass fibers (5 wt%) hybrid nanocomposites. *Iranian Polymer Journal*. 2024;33(4):511–30.
29. Li J, Sun F. The interfacial feature of thermoplastic polystyrene composite filled with nitric acid oxidized carbon fiber. *Surface Interface Analysis: An International Journal* 2009;41(3):255–8.
30. Yan C, Hao L, Xu L, Shi Y. Preparation, characterisation and processing of carbon fibre/polyamide-12 composites for selective laser sintering. *Composites Science Technology*. 2011;71(16):1834–41.
31. Fu X, Lu W, Chung D. Ozone treatment of carbon fiber for reinforcing cement. *Carbon*. 1998;36(9):1337–45.
32. Mandolfino C, Lertora E, Gambaro C. Influence of cold plasma treatment parameters on the mechanical properties of polyamide homogeneous bonded joints. *Surface Coatings Technology*. 2017;313:222–9.
33. Wolf R, Sparavigna AC. Role of plasma surface treatments on wetting and adhesion. *Engineering*. 2010;2(6):397.
34. Jones MR, Osberg KD, Macfarlane RJ, Langille MR, Mirkin CA. Templated techniques for the synthesis and assembly of plasmonic nanostructures. *Chemical Reviews*. 2011;111(6):3736–827.
35. Svanda J, Gromov M, Kalachyova Y, Postnikov P, Svorcik V, Lyutakov O. Preparation of ordered silver angular nanoparticles array in block copolymer film for surface-enhanced Raman spectroscopy. *Journal of Nanoparticle Research*. 2016;18:1–8.

36. Wang T, Jiao Y, Mi Z, Li J, Wang D, Zhao X et al. PEEK composites with polyimide sizing SCF as reinforcement: Preparation, characterization, and mechanical properties. *High Performance Polymers.* 2020;32(4):383–93.
37. Ferreira A, Pedrosa P, Lanceros-Mendez S, Machado A, Vaz F. Activation of polyethylene terephthalate using different plasma treatments. *Journal of Optoelectronics Advanced Materials.* 2010;12(7):1581.
38. Xanthos M. *Functional Fillers for Plastics.* John Wiley & Sons; 2010.
39. Alam S, Habib F, Irfan M, Iqbal W, Khalid K. Effect of orientation of glass fiber on mechanical properties of GRP composites. *Journal of the Chemical Society of Pakistan.* 2010;32(3):265–9.
40. Weitzsacker CL, Xie M, Drzal LTJS. Using XPS to investigate Fiber/Matrix chemical interactions in carbon-fiber-reinforced composites. *Surface Interface Analysis: An International Journal.* 1997;25(2):53–63.
41. Ibarra L, Macias A, Palma E. Stress-strain and stress relaxation in oxidated short carbon fiber-thermoplastic elastomer composites. *Journal of Applied Polymer Science.* 1996;61(13):2447–54.
42. Delamar M, Desarmot G, Fagebaume O, Hitmi R, Pinsonc J, Savéant J-M. Modification of carbon fiber surfaces by electrochemical reduction of aryl diazonium salts: Application to carbon epoxy composites. *Carbon.* 1997;35(6):801–7.
43. Zhang X, Pei X, Jia Q, Wang Q. Effects of carbon fiber surface treatment on the tribological properties of 2D woven carbon fabric/polyimide composites. *Applied Physics A.* 2009;95:793–9.
44. Donnet J, Brendle M, Dhami T, Bahl O. Plasma treatment effect on the surface energy of carbon and carbon fibers. *Carbon.* 1986;24(6):757–70.
45. Li R, Ye L, Mai Y-W. Application of plasma technologies in fibre-reinforced polymer composites: A review of recent developments. *Composites Part A: Applied Science Manufacturing.* 1997;28(1):73–86.
46. Tiwari S, Bijwe J, Panier S. Polyetherimide composites with gamma irradiated carbon fabric: Studies on abrasive wear. *Wear.* 2011;270(9–10):688–94.
47. Li J-Q, Huang Y-D, Fu S-Y, Yang L-H, Qu H-t, Wu G-s. Study on the surface performance of carbon fibres irradiated by γ-ray under different irradiation dose. *Applied Surface Science.* 2010;256(7):2000–4.
48. Baghaei B, Skrifvars M, Salehi M, Bashir T, Rissanen M, Nousiainen P. Novel aligned hemp fibre reinforcement for structural biocomposites: Porosity, water absorption, mechanical performances and viscoelastic behaviour. *Composites Part A: Applied Science Manufacturing.* 2014;61:1–12.
49. Chand N, Dwivedi U. Sliding wear and friction characteristics of sisal fibre reinforced polyester composites: Effect of silane coupling agent and applied load. *Polymer Composites.* 2008;29(3):280–4.
50. Dai Z, Shi F, Zhang B, Li M, Zhang Z. Effect of sizing on carbon fiber surface properties and fibers/epoxy interfacial adhesion. *Applied Surface Science.* 2011;257(15):6980–5.
51. Jang J, Yang H. The effect of surface treatment on the performance improvement of carbon fiber/polybenzoxazine composites. *Journal of Materials Science.* 2000;35:2297–303.

52. Tiwari S, Bijwe J, Panier S. Tribological studies on polyetherimide composites based on carbon fabric with optimized oxidation treatment. *Wear.* 2011;271(9–10):2252–60.
53. Aireddy H, Mishra S. Tribological behavior and mechanical properties of Bio-waste reinforced polymer matrix composites. *Journal of Metallurgy Materials Science.* 2011;53(2):139–52.
54. Schön J. Coefficient of friction and wear of a carbon fiber epoxy matrix composite. *Wear.* 2004;257(3–4):395–407.
55. El-Tayeb N. A study on the potential of sugarcane fibers/polyester composite for tribological applications. *Wear.* 2008;265(1–2):223–35.
56. Charanjit S, Lakhvir S. Synthesis and tribological characterization of Al-SiC composite prepared by mechanical alloying. *International Journal of Mechanical Engineering and Materials Science.* 2012;5:31–6.
57. Bijwe J, Indumathi J, Rajesh JJ, Fahim M. Friction and wear behavior of polyetherimide composites in various wear modes. *Wear.* 2001;249(8):715–26.
58. Wan Y, Luo H, Wang Y, Huang Y, Li Q, Zhou F et al. Friction and wear behavior of three-dimensional braided carbon fiber/epoxy composites under lubricated sliding conditions. *Journal of Materials Science.* 2005;40:4475–81.
59. Kolluri D, Satapathy B, Bijwe J, Ghosh A. Analysis of load and temperature dependence of tribo-performance of graphite filled phenolic composites. *Materials Science Engineering: A.* 2007;456(1–2):162–9.
60. Munium A, Algbory R. Wear rate behaviour of carbon/epoxy composite materials at different working conditions. *Iraqi Journal for Mechanical and Materials Engineering.* 2011;11:475–85.
61. Yousif B. Design of newly fabricated tribological machine for wear and frictional experiments under dry/wet condition. *Materials Design.* 2013;48:2–13.
62. Antaryami M. Dry sliding wear behavior of epoxy rubber dust composites. *World Academic Science Engineering Technology.* 2012;67:579–84.
63. Shivamurthy B, Prabhuswamyc M. Influence of SiO2 fillers on sliding wear resistance and mechanical properties of compression moulded glass epoxy composites. *Journal of Minerals Materials Characterization Engineering.* 2009;8(7):513.
64. Parikh HH, Gohil PP. Composites as TRIBO materials in engineering systems: Significance and applications. *Processing Techniques and Tribological Behavior of Composite Materials.* IGI Global; 2015. p. 168–91.
65. Mohan N, Natarajan S, KumareshBabu S. Investigation on two-body abrasive wear behavior of silicon carbide filled glass fabric-epoxy composites. *Journal of Minerals Materials Characterization Engineering.* 2010;9(3):231.
66. Basavarajappa S, Arun K, Davim JP. Effect of filler materials on dry sliding wear behavior of polymer matrix composites–a Taguchi approach. *Journal of Minerals Materials Characterization Engineering.* 2009;8(5):379.
67. Deo C, Acharya S. Effects of fiber content on abrasive wear of Lantana Camara fiber reinforced polymer matrix composite. 2010;17(3):219–223.
68. Chin C, Yousif B. Potential of kenaf fibres as reinforcement for tribological applications. *Wear.* 2009;267(9–10):1550–7.
69. Kranthi G, Satapathy A,Evaluation and prediction of wear response of pine wood dust filled epoxy composites using neural computation, Comput. Mater. Sci. 2010:49(3):609-614.

70. Yousif B, Lau ST, McWilliam S. Polyester composite based on betelnut fibre for tribological applications. *Tribology international*. 2010;43(1–2):503–11.
71. Mishra P, Acharya S. Anisotropy abrasive wear behavior of bagasse fiber reinforced polymer composite. *International Journal of Engineering, Science Technology*. 2010;2(11):104–112.
72. El-Sayed A, El-Sherbiny M, Abo-El-Ezz A, Aggag G. Friction and wear properties of polymeric composite materials for bearing applications. *Wear*. 1995;184(1):45–53.
73. Yousif B, Leong OB, Ong LK, Jye WK. The effect of treatment on tribo-performance of CFRP composites. *Recent Patents on Materials Science*. 2009;2(1):67–74.
74. Boopathi L, Sampath P, Mylsamy K. Influence of fiber length in the wear behaviour of borassus fruit fiber reinforced epoxy composites. *International Journal of Engineering Science Technologies*. 2012;4(9):4119–29.
75. Yousif B, El-Tayeb N. Wet adhesive wear characteristics of untreated oil palm fibre-reinforced polyester and treated oil palm fibre-reinforced polyester composites using the pin-on-disc and block-on-ring techniques. *Proceedings of the Institution of Mechanical Engineers, Part J: Journal of Engineering Tribology*. 2010;224(2):123–31.
76. Nirmal U, Hashim J, Low K. Adhesive wear and frictional performance of bamboo fibres reinforced epoxy composite. *Tribology International*. 2012;47:122–33.
77. Majhi S, Samantarai S, Acharya S. Tribological behavior of modified rice husk filled epoxy composite. *International Journal of Scientific Engineering Research*. 2012;3(6):180–4.
78. Ajay G, Navin C. Abrasive wear behaviour of bamboo powder filled polyester composite. *BioResources*. 2007;7:693–8.
79. Rao CC, Madhusudan S, Raghavendra G, Rao EV. Investigation in to wear behavior of coir fiber reinforced epoxy composites with the Taguchi method. *International Journal of Engineering Research and Applications*. 2012;2:371–4.
80. Sharma M, Rao IM, Bijwe J. Influence of fiber orientation on abrasive wear of unidirectionally reinforced carbon fiber–polyetherimide composites. *Tribology International*. 2010;43(5–6):959–64.
81. Wong K, Mohammed DS, Pickering S, Brooks R. Effect of coupling agents on reinforcing potential of recycled carbon fibre for polypropylene composite. *Composites Science Technology*. 2012;72(7):835–44.
82. Lee HS, Kim S-Y, Noh YJ, Kim SY. Design of microwave plasma and enhanced mechanical properties of thermoplastic composites reinforced with microwave plasma-treated carbon fiber fabric. *Composites Part B: Engineering*. 2014;60:621–6.
83. Han SH, Oh HJ, Kim SS. Evaluation of fiber surface treatment on the interfacial behavior of carbon fiber-reinforced polypropylene composites. *Composites Part B: Engineering*. 2014;60:98–105.
84. Cho SM, Jung H-T. Highly enhanced mechanical properties of polypropylene-long carbon fiber composites by a combined method of coupling agent and surface modification of long carbon fiber. *Macromolecular Research*. 2014;22:1066–73.

85. Liu P, Zhang S, Lu C, Yuan H. Increased interfacial adhesion between carbon fiber and poly (vinylidene fluoride) by an aqueous sizing agent. *Surface Interface Analysis*. 2016;48(13):1410–7.
86. Luo H, Xiong G, Ma C, Li D, Wan Y. Preparation and performance of long carbon fiber reinforced polyamide 6 composites injection-molded from core/shell structured pellets. *Materials Design*. 2014;64:294–300.
87. Ozkan C, Karsli NG, Aytac A, Deniz V. Short carbon fiber reinforced polycarbonate composites: Effects of different sizing materials. *Composites Part B: Engineering*. 2014;62:230–5.
88. Liu W, Zhang S, Hao L, Jiao W, Yang F, Li X et al. Properties of carbon fiber sized with poly (phthalazinone ether ketone) resin. *Journal of Applied Polymer Science*. 2013;128(6):3702–9.
89. WenBo L, Shu Z, LiFeng H, WeiCheng J, Fan Y, XiaoFei L et al. Interfacial shear strength in carbon fiber-reinforced poly (phthalazinone ether ketone) composites. *Polymer Composites*. 2013;34(11):1921–6.
90. Yu Z, Bai Y, Li Y, Wang X, Wang W, Liu J. Fiber length distribution and thermal, mechanical and morphological properties of thermally conductive polycarbonate/chopped carbon fiber composites. *Polymer International*. 2018;67(8):1137–44.
91. Qiujun L, Feng H, Ming Y, Di Y. The addition of acid treated carbon nanotube on the interfacial adhesion of carbon fiber reinforced UHMWPE composite. *Surface Interface Analysis*. 2017;49(8):717–20.
92. Li J, Xu Z. Effect of surface oxidation treatment on the tribological performance of carbon fiber reinforced polyimide composites. *Materialwissenschaft und Werkstofftechnik*. 2010;41(2):106–10.
93. Altay L, Bozaci E, Atagur M, Sever K, Tantug GS, Sarikanat M et al. The effect of atmospheric plasma treatment of recycled carbon fiber at different plasma powers on recycled carbon fiber and its polypropylene composites. *Journal of Applied Polymer Science*. 2019;136(9):47131.
94. Karsli NG, Aytac A, Akbulut M, Deniz V, Güven O. Effects of irradiated polypropylene compatibilizer on the properties of short carbon fiber reinforced polypropylene composites. *Radiation Physics Chemistry*. 2013;84:74–8.
95. Wang CC, Zhao YY, Ge HY, Qian RS. Enhanced mechanical and thermal properties of short carbon fiber reinforced polypropylene composites by graphene oxide. *Polymer Composites*. 2018;39(2):405–13.
96. Ma Y, Yan C, Xu H, Liu D, Shi P, Zhu Y et al. Enhanced interfacial properties of carbon fiber reinforced polyamide 6 composites by grafting graphene oxide onto fiber surface. *Applied Surface Science*. 2018;452:286–98.
97. Li F, Liu Y, Qu C-B, Xiao H-M, Hua Y, Sui G-X et al. Enhanced mechanical properties of short carbon fiber reinforced polyethersulfone composites by graphene oxide coating. *Polymer*. 2015;59:155–65.
98. Arao Y, Yumitori S, Suzuki H, Tanaka T, Tanaka K, Katayama T. Mechanical properties of injection-molded carbon fiber/polypropylene composites hybridized with nanofillers. *Composites Part A: Applied Science Manufacturing*. 2013;55:19–26.
99. Liu W, Li L, Zhang S, Yang F, Wang R. Mechanical properties of carbon nanotube/carbon fiber reinforced thermoplastic polymer composite. *Polymer Composites*. 2017;38(9):2001–8.

100. Luo W, Liu Q, Li Y, Zhou S, Zou H, Liang M. Enhanced mechanical and tribological properties in polyphenylene sulfide/polytetrafluoroethylene composites reinforced by short carbon fiber. *Composites Part B: Engineering*. 2016;91:579–88.
101. Vivekanandhan S, Misra M, Mohanty AK. Thermal, mechanical, and morphological investigation of injection molded poly (trimethylene terephthalate)/carbon fiber composites. *Polymer Composites*. 2012;33(11):1933–40.
102. Unterweger C, Mayrhofer T, Piana F, Duchoslav J, Stifter D, Poitzsch C et al. Impact of fiber length and fiber content on the mechanical properties and electrical conductivity of short carbon fiber reinforced polypropylene composites. *Composites Science Technology*. 2020;188:107998.
103. Kada D, Koubaa A, Tabak G, Migneault S, Garnier B, Boudenne A. Tensile properties, thermal conductivity, and thermal stability of short carbon fiber reinforced polypropylene composites. *Polymer Composites*. 2018;39(S2):E664–E70.
104. Do V-T, Nguyen-Tran H-D, Chun D-M. Effect of polypropylene on the mechanical properties and water absorption of carbon-fiber-reinforced-polyamide-6/polypropylene composite. *Composite Structures*. 2016;150:240–5.

Chapter 8

Investigations on the metallurgical profile of hard-faced mild steel with high entropy alloy AlCrMnFeNi through weld overlaying process

P. Jegannathan, R. Maanu Shree, R. Vaira Vignesh, K. Rajesh Kannan, M. Govindaraju, Thirugnasambandam G. Manivasagam, and B. Arivazhagan

8.1 INTRODUCTION

High-entropy alloys (HEAs) contain multiple primary alloying elements at nearly equiatomic ratios. They differ from traditional engineering alloys in that they are not based on a single majority component, into which minority additions are made [1]. As a result, their qualities, such as increased hardness, durability, corrosion resistance, thermal stability [2], and wear resistance, typically outperform those of ordinary alloys. Excavators, also known as diggers, are used for a variety of tasks, such as material handling, landscaping, demolition, mining projects, trenching, river dredging, and building [3]. Excavators in earthmoving and heavy mining machines that operate in hostile settings require new materials with improved strength, hardness, toughness, thermal stability [2], corrosion resistance, impact resistance, irradiation resistance, fatigue life, and wear life [4]. Excavators, for instance, are traditionally composed of mild steel. Subsequently, depending on the application and equipment cost, mild steel is hard-faced with high-chromium steel.

Mild steel is widely used in machinery and automobile manufacturing, suitable for different cutting and coating methods, and has good weldability while providing sufficient physical properties [5]. However, harsh operating conditions severely damage excavators, necessitating periodic replacement. However, the severe hardness and poor workability of HEAs limit their engineering applicability [6]. The combination of HEAs can increase the functioning of current alloys that operate in harsh environments [7]. Surprisingly, this topic has not received much attention from the scientific community. As a result, the project is focused on developing and optimizing an appropriate weld overlaying technology to join HEA with mild steel for excavators in earthmoving and heavy mining machinery.

 DOI: 10.1201/9781003713364-8

Mohammad Fereidouni et al. [8] investigated the microstructure and phases obtained by cladding $Al_xCoCrFeNi$ HEA on hot-rolled SS 304L using a tungsten inert gas heat source. The hardness of the clad zone increased steadily from 170 HV to 530 HV as the Al concentration of the HEA increased from 0.5 to 1.5. In the clad zone, a single-phase body-centred cubic (BCC) solid solution was detected. When x was less than 0.5, a single-phase face-centred cubic (FCC) solid solution formed. If x increased from 0.5 to 0.9, a dual-phase FCC and BCC crystal structure was also formed. When x was greater than 0.9, a single-phase BCC/B2 solid solution was formed. Yue et al. [9] examined the microstructure and corrosion characteristics of AISI 1045 carbon steel that has been electrospark deposited (ESD) with AlCoCrFeNi HEA. The BCC structure was observed on coated HEA. The coated HEA had an Al2O3 count that was about three times higher than the cast HEA. HEA that had been coated revealed a complete columnar crystal structure without intercellular segregation. Rathinavelu Sokkalingam et al. [10] investigated the microstructure and mechanical parameters of a dissimilar transition joint formed by conventional tungsten inert gas welding between $Al_{0.1}CoCrFeNi$-HEA and AISI304 austenitic stainless steel. According to the tensile test results, the dissimilar weld joint had a significantly higher tensile strength (590 MPa) and yield strength of 265 MPa, which was 80% greater than the substrate.

Hao Liu et al. [11] examined the microstructure and properties of AlCoCrFeNiTi HEA coating on AISI1045 steel that was fabricated by laser cladding. The AlCoCrFeNiTi HEA coating had an E_{corr} of −470 mV, which was 65% higher than that of the substrate. Similarly, the microhardness of the coating was 865 HV, ~4.5 times higher than the substrate. An average friction coefficient of the substrate was 0.469, which was ~1.6 times higher than the AlCoCrFeNiTi HEA (0.285). Ning Liu et al. [12] studied the microstructure and characteristics of CoCrFeMoNi high entropy alloy (HEA) coatings formed on the surface of SS304 using an argon tungsten arc cladding technique. HEA-coated SS304 has a maximum hardness of 585 HV (three times that of SS304). In addition, the HEA-coated SS304 had a lower volume abrasion rate (42% lower than SS304).

Qi Chao et al. [13] investigated the microstructure and characteristics of $Al_xCoCrFeNi$ HEA cladded on 253MA austenitic steel by direct laser deposition. The change in Al concentration produced an FCC phase for $x = 0.3$, a mixed FCC + BCC phase for $x = 0.6$, and a BCC phase for $x = 0.85$. The average grain size for the Al was 40.9 ± 3.7, 38.2 ± 3.1, and 32.5 ± 1.5 μm, respectively (0.3, 0.6, and 0.85). Microhardness rose as the Al content increased. As the aluminium content increased from 0.3 to 0.6 to 0.85, the microhardness values increased from 2.5 GPa to 3.2 GPa to 6.1 GPa. Olga Samoilova et al. [14] investigated the microstructure and characteristics of an $Al_{0.25}CoCrFeNiV$ HEA coating placed on stainless steel by laser cladding in an argon atmosphere. The lower the microhardness index, the higher the Fe concentration in the coating. The microhardness index (Al0.3) was 176 HV. After annealing and rolling, the microhardness was approximately 260 HV0.3. The microstructure of $Al_{0.25}CoCrFeNiV$ HEA was enhanced by accelerated heating, recrystallization, and rapid cooling under laser irradiation.

Ali Arab et al. [15] examined the microstructure and mechanical properties of AlCoCrFeNi HEA welded on Al plate using an explosive welding process. As no cracks formed, it was concluded that the optimum stand-off distance was 1 mm. Grain reinforcing and shock pressure improved the hardness near the contact. The Vickers hardness test confirmed that the HEA had a magnitude of 580 HV. Avila-Rubio et al. [16] examined the microstructure and microhardness of AlCoFeNiZn and AlCoFeNiMoTiZn HEA with inclusion Zn. According to the phase study, AlCoFeNiZn has an FCC structure, while AlCoFeNiMoTiZn has a BCC–FCC structure. Microhardness of AlCoFeNiZn was 994 HV and AlCoFeNiMoTiZn showed 865 HV. The hardness of AlCoFeNi and AlCoFeNiMoTi was 869 HV and 1225 HV, respectively. FCC-structured HEA had high ductility and toughness, whereas BCC-structured HEA had higher mechanical strength. Ge Wang et al. [17] investigated a dissimilar junction in a vacuum furnace utilizing an Ag72Cu filler material between CrMnFeCoNi HEA and SS304 alloy. The HEA-Ag72Cu interface displayed greater Mn diffusivity and solubility than the other components in the HEA. Due to the presence of Mn, greater penetration was reported at the HEA/FM interface than at the SS/FM interface. The prepared joints were found to have a strength of 459.7 MPa. By cladding, coating, and weld overlaying [18], high-entropy alloys with good tensile properties, hardness, creep and fatigue resistance, wear resistance, and corrosion resistance have been achieved in previous research and studies, owing to the four key effects of HEA [19].

Previous studies revealed that entropy alloys can be reinforced with a wide range of materials such as stainless steel, aluminium, 6082 aluminium alloy, SS304L, AISI 1045 carbon steel, and SS304 to improve the hardness, tensile strength, corrosion, and wear resistance. However, extended studies on the microstructure and mechanical properties of the weld overlay effect of AlCrMnFeNi HEA on mild steel are seldom discussed in the literature. The results were used to present a comprehensive analysis of AlCrMnFeNi high-entropy alloy-coated mild steel using autogenous tungsten inert gas (TIG) weld overlay.

8.2 MATERIALS AND METHODOLOGY

8.2.1 Design of HEA

HEA was designed based on the fraction of the constituent elements. A combination of elements with a large atomic size difference was considered for the high entropy alloy to facilitate the formation of a solid solution with high entropy [20]. The formation of solid-solution phases in multicomponent HEA plays an important role in defining their properties [21]. The Gibbs free energy and the enthalpy of mixing are important factors in determining the formation of solid-solution phases [22]. The Gibbs free energy at a specific composition is directly proportional to the free energy of mixing ($\Delta Gmix$) of the liquid phase. In the current investigation, ΔG_{mix} was calculated using Equation 8.1 [23].

$$\Delta G_{mix} = \Delta H_{mix} - T\Delta S_{mix} \tag{8.1}$$

where T is the absolute temperature, ΔH_{mix} is the enthalpy of mixing, and ΔS_{mix} is the entropy of mixing [24]. The enthalpy of mixing for high entropy alloy system with n elements was determined using Equation 8.2 and Equation 8.3.

$$\Delta H_{mix} = \sum_{i=1,\ \ i \neq j}^{n} \Omega_{ij} c_i c_j \tag{8.2}$$

$$\Omega_{ij} = 4\Delta H_{AB}^{mix} \tag{8.3}$$

where Ω_{ij} is the regular solution interaction parameter between the ith and jth elements, ci or c_j is the atomic percentage of the ith or jth component, and ΔH_{AB}^{mix} is the enthalpy of mixing of binary liquid alloys [25]. The entropy of mixing of an n-element regular solution based on Boltzmann's hypothesis is given by Equation 8.4.

$$\Delta S_{mix} = -R\sum_{i=1}^{n} c_i\, lnc_i \tag{8.4}$$

where c_i is the mole percent of component, $\sum_{i=1}^{n} c_i = 1$, and R (= 8.314 JK^{-1} mol^{-1}) is the gas constant.

Phase formation typically takes place in the vicinity of the melting temperature (Tm) of the alloy. Therefore, T_m was utilized for the entropy term ΔS_{mix}. T_m was calculated based on the rule of mixtures, as given by Equation 8.5, and a parameter Φ was introduced to forecast solid-solution formation in diverse multicomponent alloys as given by Equation 8.6.

$$T_m = \sum_{i=1}^{n} c_i \left(T_m\right)_i \tag{8.5}$$

$$\Phi = \frac{T_m \Delta S_{mix}}{\left|\Delta H_{mix}\right|} \tag{8.6}$$

However, the substantial differences in atomic size ratios could lead to serious lattice distortion within the alloy, hindering the diffusion of atoms in the matrix. The distortion lowers the transformation rate and increases the phase strain energy [26]. Consequently, atoms may segregate within the alloy, resulting in an increase in free energy and potentially leading to the formation of nanocrystalline and amorphous structures. Conversely, significant differences in atomic size could also cause sluggish diffusion of atoms in the matrix, exacerbating lattice distortion and increasing phase strain energy, thereby impacting solid-solution stability [27]. Hence, the size effect stemming from atomic disparities in an *n*-element alloy was delineated using δ. The parameter δ was calculated using Equation 8.7 and Equation 8.8.

$$\delta = \sqrt{\sum_{i=1}^{n} c_i \left(1 - \frac{r_i}{\underline{r}}\right)^2} \tag{8.7}$$

$$\underline{r} = \sum_{i=1}^{n} c_i r_i \tag{8.8}$$

where $\underline{r}$ is the average atomic radius and r_i is the atomic radius of element.

8.2.3 Numerical evaluation of solid-phase fraction

A computational thermodynamic simulation aimed at understanding the solidification behaviour of the AlCrMnFeNi HEA. The weight fraction of the elements at various temperatures, solid-phase fraction, and microstructural evolution of the fabricated AlCrMnFeNi HEA were theoretically determined using the computing environment MATLAB 2024® coupled with a professional simulation software.

8.2.4 Materials

Commercially available mild steel (substrate) and self-prepared AlCrMnFeNi high-entropy alloys were used as base materials. The constituent powders of the entropy alloy (Al, Cr, Mn, Fe, and Ni) had an average particle size of 30 μm with a purity of ~99.9%. A mild steel plate was cut into specimens of dimensions 10 mm × 10 mm × 5 mm.

8.2.5 Fabrication of high entropy alloy

The AlCrMnFeNi HEA specimens were fabricated using spark plasma sintering [28]. The metal powders (Al, Cr, Mn, Fe, and Ni) in equimolar proportions were homogeneously mixed using a high-energy ball-milling unit. The constituent powders were added to a stainless steel container with 50 alumina ceramic balls and subsequently rotated at 750 rpm for 60 minutes. The ball-milled powders were then consolidated by spark plasma sintering in a 25 mm inner-diameter graphite die. The powders were heated to 1200°C at a heating rate of 50°C/min, and then were sintered at 1200°C for 20 minutes.

8.2.6 Weld overlaying of HEA over mild steel

TIG welding with argon shielding gas was utilized for the surface melting and alloying of HEA over a mild steel substrate [1]. The fabricated AlCrMnFeNi HEA specimen was placed on a mild steel substrate. The TIG torch was moved over the HEA alloy with a potential difference of 2.2 V, welding current of 120 A, and welding speed of 18 mm/s. The flow rate of argon gas was maintained at 9 L/min. The conjoined workpiece was subjected to metallurgical characterization.

8.2.7 Microstructure

The specimens for the microstructure analysis were prepared from the transverse section of the weld-overlay workpiece [29]. The specimen preparation followed the guidelines of ASTM E3-11, including polishing with standard metallographic sheets (600, 800, 1000, and 2000 grit) and fine polishing in a diamond suspension. The microstructure was observed using high-resolution scanning electron microscopy.

8.2.8 Microhardness

Microhardness was analyzed on the HEA, at the interface of the HEA and substrate, and on the substrate. The specimens were prepared in accordance with the standard ASTM E3-11 [30]. Vickers microhardness was measured using a microhardness tester (make: UHL, model: VMHT 104), as outlined by the standard ASTM E92. The specimen was indented using a diamond indenter at an axial load of 500 g for 15 seconds.

8.2.9 Tribological characterization

The wear rate and coefficient of friction of the specimens were determined using a pin-on-disc tribometer (make: Ducom, model: TR20LE). Tribotest specimens were prepared by mounting a polished specimen with dimensions of 10 mm × 10 mm × 10 mm in a hollow cylindrical rod [31]. The specimens were cleaned with acetone and weighed using a precision weighing balance with a readability of 0.0001 g. The wear test was conducted in accordance with the ASTM G99–95a standard [32]. The specimen was slid against an Al_2O_3 ceramic disc under an axial load of 10 N at a sliding velocity of 0.20 ms^{-1} for 360 m.

8.2.10 Instrumental characterization

The microstructure and surface morphologies of the specimens subjected to various tests were observed using a field-emission scanning electron microscope (FE-SEM; make: Zeiss Sigma, model: Gemini 300). The images were obtained at electron acceleration voltages of 5 kV and 10 kV at various magnifications. The elemental composition and elemental map of the corrosion products and corroded regions in the specimens were acquired using energy-dispersive x-ray spectroscopy (EDS; make: oxford Instruments, model: Ultim Max) at an electron high tension of 20 kV. Phase analysis was carried out using x-ray diffractometry (XRD; make: Rigaku, model: Ultima 4). X-ray diffraction graphs were obtained in the continuous scanning mode using Cu Kα radiation at a scan rate of 0.5°/min between 10° and 90°.

8.3 RESULTS AND DISCUSSION

8.3.1 Design of high entropy alloy

A negative $\Delta Hmix$ tends to facilitate the formation of intermetallic compounds, indicating strong elemental binding, while a positive ΔH_{mix} suggests limited miscibility in the liquid phase, leading to element segregation or a miscibility gap. In multicomponent alloys, ΔH_{mix} exerts a similar influence, promoting intermetallic compound formation with negative values and reducing liquid alloy miscibility with positive values, resulting in element segregation. Solid solution formation is favoured when ΔH_{mix} approaches zero, allowing for random element distribution and stable solid solutions, consistent with prior findings [33].

Additionally, the contribution of $\Delta Smix$, particularly in high-entropy alloy systems where ΔS_{mix} is consistently positive, warrants consideration. High ΔS_{mix} increases disorder, lowering ΔG_{mix}, and facilitating random solid solution formation, which is more stable than alternative phases during solidification. Moreover, as temperature rises, the influence of ΔS_{mix} becomes more pronounced, further promoting solid solution formation, potentially preceding ordered phases [34]. Hence, $T\Delta Smix$ becomes the prime factor for solid solution formation at high temperatures. Consequently, solid solutions may form before ordered phases, typically near the melting temperature, irrespective of solid-state phase transitions [23].

If the value of Φ is positive, then $\Phi = 1$ is suggested as a critical threshold for solid solution formation. When $\Phi > 1$, the contribution of $T\Delta S_{mix}$ surpasses that of $\Delta Hmix$ in solid solution formation, leading to multicomponent high-entropy alloys primarily composed of solid solutions. Conversely, if $\Phi \leq 1$, ΔH_{mix} predominates in the free energy, resulting in the formation of intermetallic compounds and segregations before solid solution phases in multicomponent HEAs [35]. The calculated ΔS_{mix} and ΔH_{mix} for the high entropy alloy was determined to be 14.2698 kJ/mol and 3.4783 kJ/mol. Substituting in Equation 8.6, the parameter Φ was found to be greater than 1. Hence, the HEA would be a solid solution without any intermetallic compounds [23]. Also, δ tends to minimal value indicating the limited possibility for the segregation of elements.

The simulation indicated that AlCrMnFeNi HEA would be composed of FCC and BCC phases. AlCrMnFeNi primarily exhibited the FCC crystal structure. Figure 8.1a indicates the temperature range over which the AlCrMnFeNi HEA transitions from a completely liquid to a completely solid state. The graph shows a relatively narrow solidification range, indicating AlCrMnFeNi HEA solidifies over a small temperature interval. The presence of separate lines for BCC and FCC phases implies that the AlCrMnFeNi HEA undergoes solidification with the formation of distinct crystal structures. The FCC phase appears to start solidifying first. This suggests that the FCC phase is more stable at higher temperatures within the solidification range of AlCrMnFeNi HEA. The full solidification occurred at a temperature just above 1200°C. The lines do not intersect at any point, which indicated that the BCC and FCC phases coexist in the alloy as the temperature decreased. The final microstructure of the alloy, which is a mix

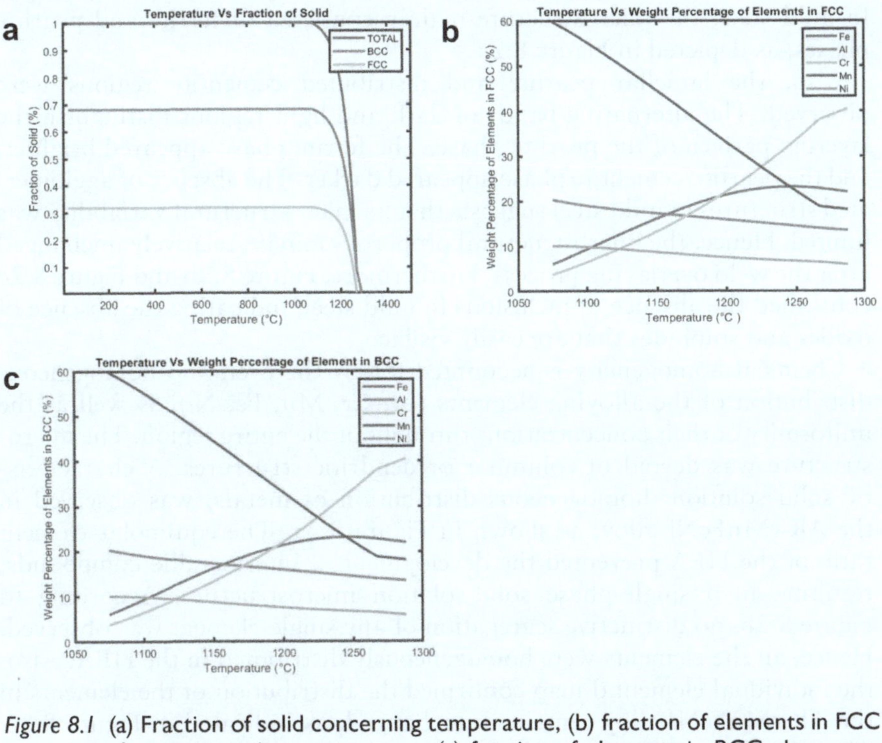

Figure 8.1 (a) Fraction of solid concerning temperature, (b) fraction of elements in FCC phase concerning temperature, (c) fraction of elements in BCC phase concerning temperature.

of BCC and FCC phases, will have implications for its mechanical properties, such as strength, ductility, and toughness [36].

Figure 8.1b shows the weight percentages of the elements in the FCC phase of the AlCrMnFeNi HEA system concerning temperatures. The fraction of Mn increased in the FCC phase as the temperature decreased. However, the fraction of Fe in FCC decreased from 22% to 6%, showing a significant reduction in weight percentage with respect to the temperature [25]. Figure 8.1c represents the temperature and composition range of the BCC phase in AlCrMnFeNi HEA. The size and atomic radii of the elements will influence the stability of the BCC lattice. However, processing conditions (cooling rate) can significantly influence the formation of the BCC phase.

8.3.2 Microstructure

The investigation focused on the microstructure evolution model of a standard joint between AlCrMnFeNi HEA and mild steel. Additionally, various elemental mapping locations were utilized to distinguish the interface between the HEA and mild steel. The coated mild steel cross-section was subject to elemental mapping, and the results are presented in Figure 8.2. In mild steel, the microstructure mainly consisted of ferrite and pearlite phases, as depicted in Figure 8.2a.

Also, the lamellar pearlite and distributed cementite regions were observed. The alternating bands of dark and light regions distinguish the layering pattern of the pearlite phase. The ferrite phase appeared brighter, and the pearlite/cementite phase appeared darker. The absence of agglomerated structures in mild steel suggests that its microstructural variability was limited. Hence, the microstructural properties remain relatively unchanged after the weld overlaying process. Furthermore, Figure 8.2b and Figure 8.2c confirmed the absence of inclusions in mild steel, indicating the absence of oxides and sulphides that are easily visible.

Chemical homogeneity is accounted for by the even and homogeneous distribution of the alloying elements (Al, Cr, Mn, Fe, Ni), as well as the uniformity of their concentrations throughout the entire region. The microstructure was devoid of columnar or dendritic structures. A characteristic solid solution (homogeneous distribution of metals) was observed in the AlCrMnFeNi alloy, as shown in Figure 8.3a. The equimolar element ratio of the HEA prevented the development of intermetallic compounds, resulting in a single-phase solid-solution microstructure. According to Figure 8.3b, no distinctive segregation of any single element was observed. Hence, all the elements were homogeneously distributed in the HEA. Also, the individual elemental map confirmed the distribution of the elements in the AlCrMnFeNi alloy remained unchanged, as indicated in Figure 8.3c. Also, the AlCrMnFeNi alloy lacks any surface flaws like pores, fissures, or oxide layers.

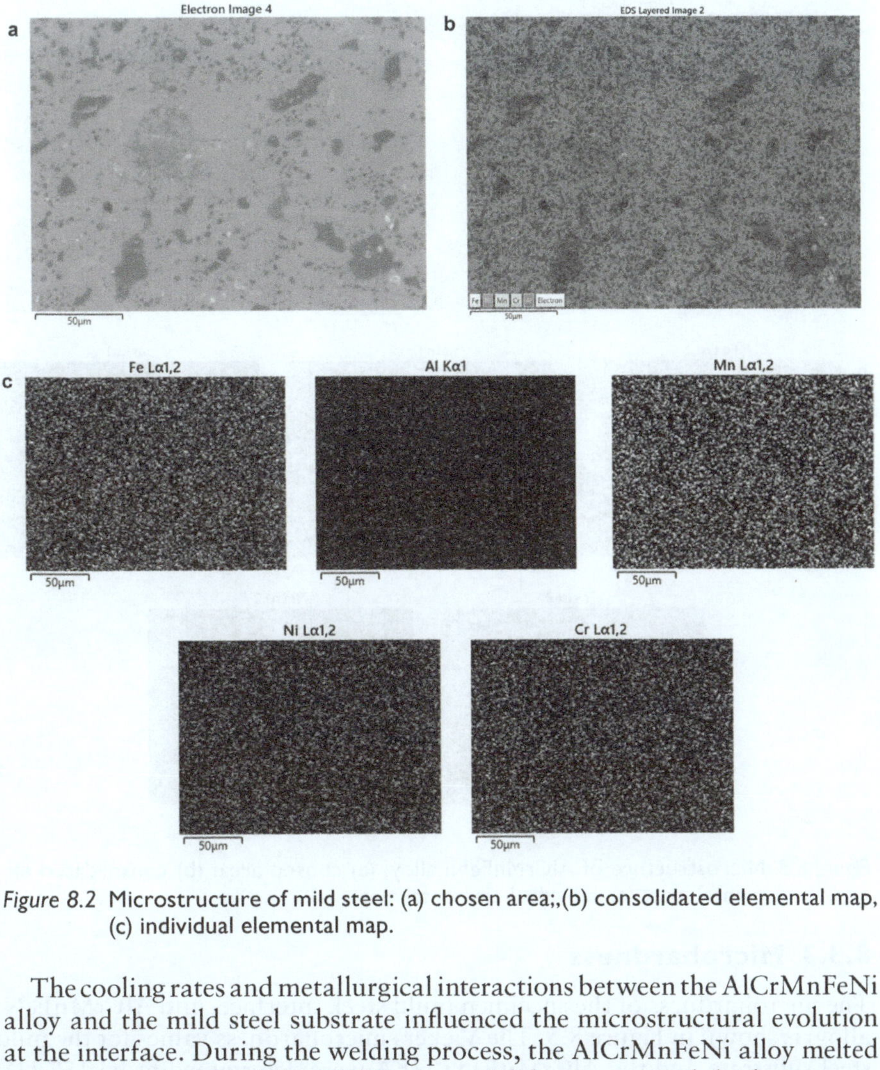

Figure 8.2 Microstructure of mild steel: (a) chosen area;,(b) consolidated elemental map, (c) individual elemental map.

The cooling rates and metallurgical interactions between the AlCrMnFeNi alloy and the mild steel substrate influenced the microstructural evolution at the interface. During the welding process, the AlCrMnFeNi alloy melted and conjoined with mild steel. The fast solidification caused the fusion zone to typically have a fine-grained microstructure. The area surrounding the fusion zone that reached high temperatures during welding but did not melt is known as the heat affected zone (HAZ) [37]. EDS analysis revealed a uniform distribution of all elements in the heat-affected zone, as shown in Figure 8.4. Hence, the interface was devoid of any specific element segregation within distinct zones. In the interface region, the mixing or interdiffusion of elements led to the formation of a transition area with varying compositions.

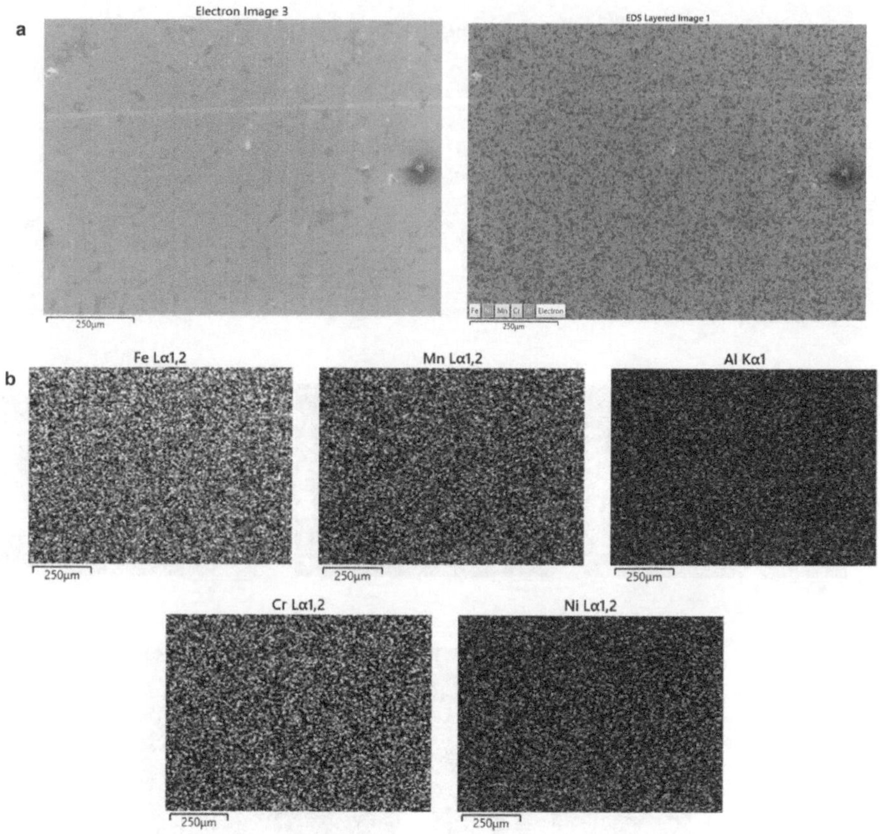

Figure 8.3 Microstructure of AlCrMnFeNi alloy: (a) chosen area, (b) consolidated elemental map, (c) individual elemental map.

8.3.3 Microhardness

The microhardness of the specimen (mild steel, interface, and AlCrMnFeNi alloy) is shown in Figure 8.5. The Vickers microhardness values for the mild steel substrate and the AlCrMnFeNi HEA were determined to be 257 HV and 467 HV, respectively, confirming the higher hardness of the HEA as reported in the existing literature. The application of the HEA weld overlay on the mild steel substrate resulted in an increased hardness, surpassing the hardness of the mild steel (~81%). The presence of the HEA in the weld overlayed zone was identified as the contributing factor to this hardness improvement. Upon analyzing the interface, a Vickers microhardness value of 284 HV was recorded, indicating the confluence of both mild steel and the high entropy alloy. However, the relatively low hardness value at the interface raised the possibility of softening or dilution of the high entropy alloy in this region.

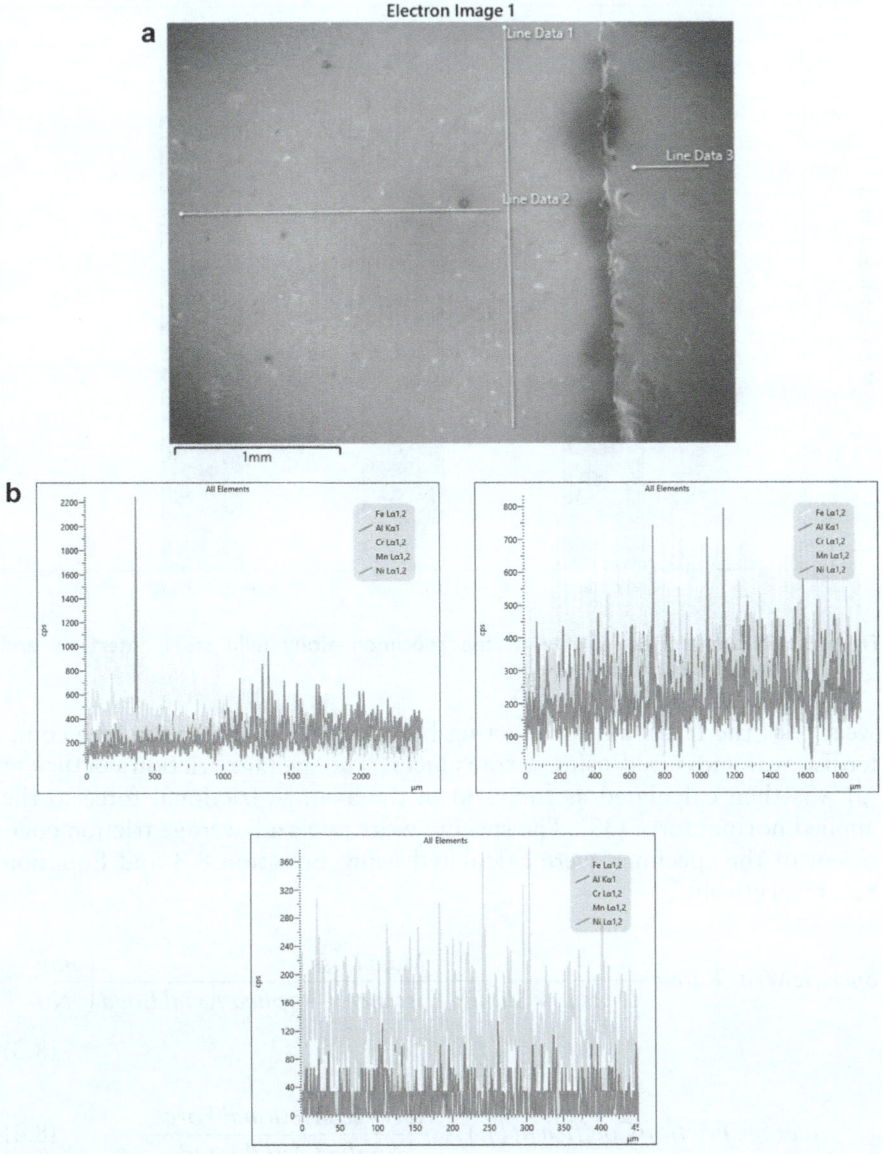

Figure 8.4 Microstructure of the interface between mild steel and AlCrMnFeNi alloy: (a) chosen region, (b) elemental concentration in the lines drawn on the chosen region.

8.3.4 Wear rate and friction coefficient

The bonded specimen was subjected to a wear test by sliding the AlCrMnFeNi HEA-mild steel against the counterdisc under controlled conditions of axial load, sliding velocity, and sliding distance. During the

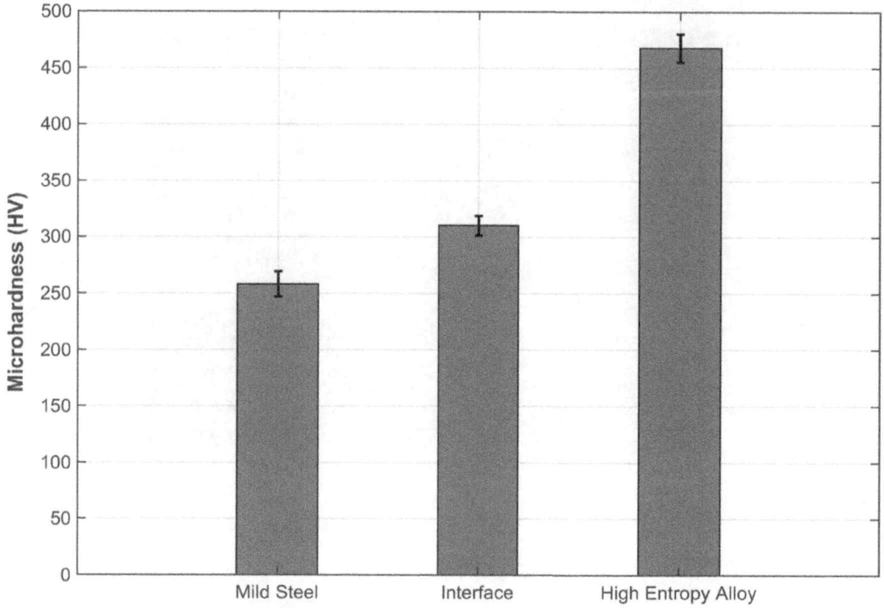

Figure 8.5 Microhardness profile of the specimen along mild steel, interface, and AlCrMnFeNi alloy.

wear test, the frictional force acting between the specimen and the counterdisc was measured using a transducer. The average friction coefficient (μ) was then calculated as the ratio of the average frictional force to the applied normal force [38]. The specific wear rate and average friction coefficient of the specimen were calculated using Equation 8.3 and Equation 8.4, respectively.

$$Specific\,Wear\,Rate = \frac{Mass\,Loss}{Density \times Sliding\,Distance \times Applied\,Axial\,Load}\left(\frac{mm^3}{Nm}\right) \tag{8.3}$$

$$Average\,Friction\,Coefficient\,(\mu) = \frac{Average\,Frictional\,Force}{Applied\,Axial\,Load} \tag{8.4}$$

The specific wear rate and average friction coefficient of AlCrMnFeNi HEA were 18.12×10^{-5} mm^3/Nm and 0.35, respectively. The surface morphology and elemental map of the specimen are shown in Figure 8.6. Under the sliding conditions, AlCrMnFeNi HEA was exposed to a thermal and mechanical environment that might induce oxidation. The distribution of elements, including oxygen, across the surface of the worn AlCrMnFeNi HEA was

Figure 8.6 Worn morphology of AlCrMnFeNi alloy: (a) chosen area, (b) consolidated elemental map, (c) individual elemental map.

homogeneous. Hence, the presence of oxide layers and regions with varying levels of oxidation was evident. The oxide scales were made up of a combination of oxides from the AlCrMnFeNi component elements. The presence of these elements, as well as their propensity to produce stable oxides, influences the protective character of the oxide layer. The well-formed, uniform, and adherent oxide layer acted as a protective barrier, preventing direct contact between the AlCrMnFeNi HEA and the countersurface, thus reducing wear rates. Also, the oxide layer reduced friction by acting as a solid lubricant.

8.4 CONCLUSION

The AlCrMnFeNi HEA and mild steel were successfully conjoined using the weld overlaying process. The microstructural evolution, microhardness, and tribological characteristics of the specimens were analyzed. The results demonstrated the following:

- The welding process produced a stable and homogeneous microstructure in both materials, with limited changes in the microstructure of mild steel.
- The high hardness of HEA and the formation of an adherent oxide layer contributed to improved wear resistance and friction reduction during sliding conditions. However, the interface region between the HEA and mild steel exhibited lower hardness, suggesting the need for further optimization to prevent dilution of elements.

DECLARATIONS

Funding

The authors are thankful to the University Grants Commission, Department of Atomic Energy, Consortium for Research, Government of India for providing financial assistance to carry out the research work "Development of High Entropy Alloy (HEA) hard-facing material for nuclear applications" through the project vide 'CRS/2022-23/1104'.

Conflicts of interest/competing interests

The authors declare that there is no conflict of interest or competing interests for the research work.

Availability of data and material (data transparency)

All data generated or analyzed during this study are included in this published article (and its supplementary information files).

Code availability, ethics approval, consent to participate, consent for publication

Not applicable.

REFERENCES

1. Hsieh, K.-C., et al., The microstructure and phase equilibrium of new high performance high-entropy alloys. *Journal of Alloys and Compounds*, 2009. **483**(1): p. 209–212.
2. Chen, J., et al., A review on fundamental of high entropy alloys with promising high–temperature properties. *Journal of Alloys and Compounds*, 2018. **760**: p. 15–30.
3. Tsai, M.-H. and J.-W. Yeh, High-entropy alloys: A critical review. *Materials Research Letters*, 2014. **2**(3): p. 107–123.
4. Fernández, J.E., et al., Materials selection to excavator teeth in mining industry. *Wear*, 2001. **250**(1): p. 11–18.
5. Li, L., et al., Effect of corrosion and hydrogen embrittlement on microstructure and mechanical properties of mild steel. *Construction and Building Materials*, 2018. **170**: p. 78–90.
6. Li, S., et al., Weldability of high entropy alloys: Microstructure, mechanical property, and corrosion resistance. *Journal of Manufacturing Processes*, 2023. **99**: p. 209–229.
7. Yeh, J.-W., et al., High-entropy coatings, in *High-Entropy Alloys: Fundamentals and Applications*, M.C. Gao, et al., Editors. 2016, Springer International Publishing: Cham. p. 469–491.
8. Fereidouni, M., M. Sarkari Khorrami, and M. Heydarzadeh Sohi, Liquid phase cladding of Al_xCoCrFeNi high entropy alloys on AISI 304L stainless steel. *Surface and Coatings Technology*, 2020. **402**: p. 126331.
9. Li, Q.H., et al., Microstructure and corrosion properties of AlCoCrFeNi high entropy alloy coatings deposited on AISI 1045 steel by the electrospark process. *Metallurgical and Materials Transactions* A, 2013. **44**(4): p. 1767–1778.
10. Sokkalingam, R., et al., Dissimilar welding of Al0.1CoCrFeNi high-entropy alloy and AISI304 stainless steel. *Journal of Materials Research*, 2019. **34**(15): p. 2683–2694.
11. Liu, H., et al., Microstructure and properties of AlCoCrFeNiTi high-entropy alloy coating on AISI1045 steel fabricated by laser cladding. *Journal of Materials Engineering and Performance*, 2019. **28**(3): p. 1544–1552.
12. Liu, N., et al., Optimizing 304 stainless steel surface performance with CoCrFeMoNi high-entropy alloy coating via gas tungsten arc cladding. *Journal of Materials Engineering and Performance*, 2023. **32**, pages 9114–9120.
13. Chao, Q., et al., Direct laser deposition cladding of AlxCoCrFeNi high entropy alloys on a high-temperature stainless steel. *Surface and Coatings Technology*, 2017. **332**: p. 440–451.
14. Samoilova, O., et al., Al0.25CoCrFeNiV High entropy alloy coating deposited by laser cladding on stainless steel. *Materials*, 2022. **15**, DOI: 10.3390/ma15207058.
15. Arab, A., et al., Joining AlCoCrFeNi high entropy alloys and Al-6061 by explosive welding method. *Vacuum*, 2020. **174**: p. 109221.
16. Avila-Rubio, M.A., et al., Microstructure and microhardness of high entropy alloys with Zn addition: AlCoFeNiZn and AlCoFeNiMoTiZn. *Advanced Powder Technology*, 2021. **32**(12): p. 4687–4696.

17. Wang, G., et al., Mechanical properties and microstructure evolution of CrMnFeCoNi HEA/304 SS dissimilar brazing joints. *Journal of Alloys and Compounds*, 2020. **829**: p. 154520.
18. Zhang, Y., et al., Research and development of welding methods and welding mechanism of high-entropy alloys: A review. *Materials Today Communications*, 2021. **28**: p. 102503.
19. Zhang, Y., et al., Microstructures and properties of high-entropy alloys. *Progress in Materials Science*, 2014. **61**: p. 1–93.
20. Gao, M.C., et al., Design of refractory high-entropy alloys. *JOM*, 2015. **67**(11): p. 2653–2669.
21. Shafiei, A., Design of eutectic high entropy alloys. *Metallurgical and Materials Transactions A*, 2022. **53**(12): p. 4349–4361.
22. Gao, M.C., et al., Thermodynamics of concentrated solid solution alloys. *Current Opinion in Solid State and Materials Science*, 2017. **21**(5): p. 238–251.
23. Yang, X. and Y. Zhang, Prediction of high-entropy stabilized solid-solution in multi-component alloys. *Materials Chemistry and Physics*, 2012. **132**(2): p. 233–238.
24. Lu, Y., et al., A new strategy to design eutectic high-entropy alloys using mixing enthalpy. *Intermetallics*, 2017. **91**: p. 124–128.
25. Ye, Y.F., et al., Design of high entropy alloys: A single-parameter thermodynamic rule. *Scripta Materialia*, 2015. **104**: p. 53–55.
26. Tan, Y., et al., Design of high-entropy alloys with a single solid-solution phase: Average properties vs. their variances. *Journal of Alloys and Compounds*, 2018. **742**: p. 430–441.
27. Zhang, Y., X. Yang, and P.K. Liaw, Alloy design and properties optimization of high-entropy alloys. *JOM*, 2012. **64**(7): p. 830–838.
28. Yang, Y., et al., Microstructure and properties of FeCoCrNiMoSix high-entropy alloys fabricated by spark plasma sintering. *Journal of Alloys and Compounds*, 2021. **884**: p. 161070.
29. Li, H., et al., Effect of tool profile on wear and mechanical behaviors of CeO2 and ZrO2-reinforced hybrid magnesium matrix composite developed via FSP technique. *Journal of Manufacturing Processes*, 2023. **94**: p. 297–315.
30. Vaira Vignesh, R. and R. Padmanaban, Influence of friction stir processing parameters on the wear resistance of aluminium alloy AA5083. *Materials Today: Proceedings*, 2018. **5**(2, Part 2): p. 7437–7446.
31. Arjun, K.B., et al., Effect of load, sliding distance and sliding velocity on the wear properties of aluminum alloy AA5052. *IOP Conference Series: Materials Science and Engineering*, 2019. **577**(1): p. 012016.
32. Luo, J., et al., Enhanced mechanical and tribological properties of AA6061/CeO2 composite fabricated by friction stir processing. *Materials Letters*, 2022. **318**: p. 132210.
33. Tancret, F., et al., Designing high entropy alloys employing thermodynamics and Gaussian process statistical analysis. *Materials & Design*, 2017. **115**: p. 486–497.
34. Senkov, O.N. and D.B. Miracle, A new thermodynamic parameter to predict formation of solid solution or intermetallic phases in high entropy alloys. *Journal of Alloys and Compounds*, 2016. **658**: p. 603–607.

35. Troparevsky, M.C., et al., Criteria for predicting the formation of single-phase high-entropy alloys. *Physical Review X*, 2015. 5(1): p. 011041.
36. Zhang, F., et al., An understanding of high entropy alloys from phase diagram calculations. *Calphad*, 2014. **45**: p. 1–10.
37. Li, J., et al., Welding of high entropy alloys: Progresses, challenges and perspectives. *Journal of Manufacturing Processes*, 2021. **68**: p. 293–331.
38. Zum Gahr, K.H., *4* – Tribological aspects of microsystems, in *Handbook of Sensors and Actuators*, T. Fukuda and W. Menz, Editors. 1998, Elsevier Science B.V. p. 83–113.

Chapter 9

Surface treatment of bioimplant materials using electrical discharge machining (EDM)

S. Senthil Kumar, R. Muralikannan, S. Rathinavel, and T.S. Senthilkumar

9.1 INTRODUCTION

In recent years, the science of bioimplant materials has made major advances, transforming medical treatments and increasing the quality of life for millions of patients worldwide. Bioimplants are artificial devices or tissues that are designed to replace, support, or increase the function of organic structures in the human body that are damaged or absent.[1] These implants are essential in a variety of medical fields, including orthopedics, dentistry, cardiology, neurology, and ophthalmology. The capacity of bioimplants to merge effortlessly with host tissue, assuring biocompatibility, osseointegration, and long-term stability, is critical to their success. The surface properties of bioimplant materials are critical in defining how they interact with the biological environment.[2] Traditional manufacturing procedures, on the other hand, may not always produce surfaces with the appropriate characteristics required for good bioimplant performance.

9.1.1 Background

Researchers and medical experts have been investigating various surface treatment strategies to improve the performance of bioimplant materials in recent years. These treatments are designed to increase the surface roughness, biocompatibility, and corrosion resistance of the implants, resulting in improved osseointegration and a lower chance of implant failure. EDM is a nontraditional machining technology that has been widely employed in the industrial industry to make detailed forms and patterns on metals and alloys.[3] EDM is the controlled erosion of material caused by sparks generated between a tool electrode and the workpiece in a dielectric medium. The method is well known for its accuracy and capacity to operate with difficult-to-machine materials.

DOI: 10.1201/9781003713364-9

9.1.2 Motivation

Several issues motivate the investigation of EDM as a surface treatment technology for bioimplant materials:

Surface enhancement: Traditional surface treatments may not give the desired surface qualities required for bioimplants. EDM is a unique method for precisely modifying the surface of bioimplant materials

Biocompatibility improvement: Improving the surface properties of bioimplants can greatly improve biocompatibility, lowering the likelihood of adverse responses or implant rejection.

Osseointegration promotion: Using EDM to create certain surface textures may increase bone cell adhesion and proliferation, resulting in improved osseointegration and long-term implant durability.

Enhancement of corrosion resistance: EDM can change the surface composition of bioimplant materials, potentially lowering corrosion and prolonging implant lifespan.

Interdisciplinary potential: The combination of EDM technology and bioimplant materials gives great prospects for collaboration between technical and medical specialists, driving novel solutions.

9.2 OBJECTIVES

The following are the key goals of this book chapter:

- To provide a thorough understanding of EDM and its variations as a surface treatment technology.
- Investigate the use of EDM to improve the surface characteristics of bioimplant materials.
- To examine the effects of EDM surface treatment on bioimplant biocompatibility, osseointegration, and corrosion resistance.
- To provide case studies demonstrating the successful application of EDM in the surface treatment of bioimplant materials.
- To talk about the biological evaluation and safety of EDM-treated bioimplants.
- To explore future views, problems, and research initiatives in the realm of surface treatment of EDM-based bioimplant materials.

9.2.1 Scope

This chapter will largely concentrate on the use of EDM for the surface treatment of bioimplant materials, including but not limited to:

- Titanium and its alloys
- Stainless steel
- Cobalt-chromium-molybdenum alloys

The chapter will cover core EDM principles, bioimplant material treatment adjustments, and the consequences on biocompatibility, osseointegration, and corrosion resistance. It will also include case studies, biological evaluations, safety concerns, and regulatory issues concerning the use of EDM in bioimplant materials. While this chapter strives to give a comprehensive assessment of the subject, it may not include all emerging breakthroughs or recent research in the field because it is based on information accessible up to the September 2021 knowledge cutoff date. Nonetheless, the chapter will be a significant resource for researchers, engineers, and medical professionals interested in the potential of EDM in bioimplant material surface treatment and its implications for enhancing patient outcomes and advancing healthcare.

9.3 CHALLENGES WITH BIOIMPLANT MATERIALS AND THEIR SURFACES

9.3.1 Materials and overview of bioimplants

Bioimplants are specialized medical devices or tissues that are meant to be implanted within the human body in order to replace, sustain, or improve the function of damaged or missing biological structures. These implants have dramatically improved patient outcomes and quality of life in a variety of medical professions. Orthopedic implants (e.g., joint replacements, bone plates, and screws),[4] dental implants,[5] cardiovascular stents,[6] artificial heart valves,[7] and neurological implants (e.g., cochlear implants and deep brain stimulators) are examples of common bioimplants.

Bioimplant materials are carefully chosen to suit specific needs such as mechanical strength, biocompatibility, corrosion resistance, and fatigue resistance. Bioimplants are commonly made from the following materials:

Titanium and its alloys: Because of their great biocompatibility, high strength-to-weight ratio, corrosion resistance, and ability to osseointegrate with bone, titanium and titanium alloys are frequently employed.[8]

Stainless steel: Because of their good mechanical qualities and biocompatibility, stainless steel alloys are extensively utilized in orthopedic and dental implants.[9]

Cobalt-chromium-molybdenum alloys: Because of their excellent strength and wear resistance, these alloys are frequently employed in orthopedic implants, particularly in load-bearing applications.[10]

Polymer-based materials: Biocompatible polymers are used to make some bioimplants such as soft tissue replacements and medication delivery systems.[11]

9.4 ESSENTIAL SURFACE QUALITIES FOR BIOIMPLANT SUCCESS

The ability of bioimplants to integrate into host tissue and execute their intended function without producing adverse reactions or consequences is critical to their effectiveness. The surface characteristics of bioimplant materials are critical in achieving these goals. Some of the surface qualities that are critical for bioimplant success are as follows:

Biocompatibility: When in touch with biological tissues, the surface should be biocompatible, which means it should not evoke an immune response or harmful effects.

Osseointegration: A surface that promotes osseointegration is required for implants intended to combine with bone. It encourages bone cell adherence and development on the implant's surface.[12]

Corrosion resistance: To retain structural integrity and prevent the discharge of toxic metallic ions, bioimplants must endure the corrosive environment of the human body.

Surface roughness: Controlled surface roughness can improve cell adhesion, proliferation, and tissue ingrowth, increasing the implant's stability and durability.

Wettability: The surface wettability of the implant can influence cell attachment and protein adsorption, hence altering biocompatibility and tissue response.[13]

Wear resistance: Implants that are prone to wear or friction, such as joint replacements, necessitate surfaces with strong wear resistance to reduce debris production and associated inflammatory reactions.

9.5 BIOIMPLANT SURFACES-RELATED DIFFICULTIES

Despite major breakthroughs in bioimplant technology, several problems remain concerning the surfaces of these implants:

Biocompatibility and immunogenicity: It is difficult to achieve good biocompatibility without inducing an immunological response or undesirable responses. The implant may be perceived as a foreign item by the body's immune system, prompting it to reject it.

Infection risk: The surface of the implant might be a site for bacterial adhesion and biofilm formation, leading to implant-associated infections.[14]

Interface between the implant and the host: It is critical for long-term implant success to ensure a robust and stable contact between the implant and the host tissue. Implant loosening or failure might result from poor integration.

Corrosion and metallic ion release: Metal or alloy bioimplants may corrode in the body, resulting in the release of metallic ions that might trigger localized tissue responses.[15]

Surface wear and particle generation: Wear and friction at the implant's surface can result in debris, which can cause inflammation and implant failure.[16]

Durability of surface coating: Some implants are coated with particular materials or substances to improve their performance. It is critical to ensure the endurance and stability of these coatings over time.

Addressing these issues and refining the surface qualities of bioimplant materials is critical for advancing implant technology, improving patient outcomes, and ensuring bioimplant treatments' long-term success. EDM is a viable method for addressing some of these issues since it allows for precise and controlled alterations of bioimplant surfaces to improve biocompatibility and performance.

9.6 ELECTRICAL DISCHARGE MACHINING (EDM)

9.6.1 Fundamentals of EDM

EDM is a nontraditional machining technology that erodes material from a workpiece using electrical sparks. It was invented in the 1940s and has since become a popular method for machining complex forms and hard materials that are difficult to produce using traditional methods.[17] The following are the fundamental principles of EDM:

Spark generation: EDM operates on the electrical discharge concept between a tool electrode (usually made of copper or graphite) and the workpiece. A high-energy electrical spark is generated in the narrow gap between the tool and the workpiece when a voltage is introduced between them.

Dielectric medium: The spark happens in a dielectric medium that surrounds the tool and workpiece (often deionized water or specialist EDM lubricants). The dielectric medium functions as an insulator, limiting continuous electrical arcing and making eroded material removal easier.

Erosion process: A high temperature is generated at the point of contact between the tool and the workpiece during the spark. This high temperature melts and vaporizes the substance, causing it to be removed as tiny particles.

Flush and filtration: The dielectric medium is continuously flushed through the gap between the tool and the workpiece to maintain stable machining conditions and remove eroded particles. Filtration systems aid in the removal of debris from dielectric fluid, which ensures machining precision and surface smoothness.

Controlled tool movement: The tool electrode is computer controlled and follows a specified toolpath, allowing for precision machining of the workpiece's desired shape.

9.6.2 EDM process types

The EDM method has various versions, each suited for unique purposes and requirements. The following are the primary types of EDM processes:

Wire EDM (WEDM): A thin, electrically conducting wire acts as the tool electrode in wire EDM. The wire is continually passed into the workpiece, and the spark erosion cuts the material with high precision to create complicated shapes.[18]

Sinker EDM (SEDM): The tool electrode in sinker EDM is typically a three-dimensional negative of the desired shape. It is immersed in dielectric fluid and moved repeatedly close to the workpiece to generate sparks and degrade the material.[19]

Hole drilling EDM (HEDM): This method is used to create small, accurate holes in hard materials. The repetitive impact of electrical discharges on a revolving electrode creates a hole.[20]

Small hole EDM (SHEDM): Like HEDM, SHEDM makes deep, accurate holes in conductive materials. It makes high-precision holes with a tubular electrode.[21]

9.7 EDM PARAMETERS AND THEIR IMPACTS ON SURFACE INTEGRITY AND MATERIAL REMOVAL RATE

EDM performance is governed by a variety of parameters that can be changed to achieve specific machining objectives.[22] The following are some major EDM parameters and their effects on material removal rate and surface integrity:

Pulse current and duration: Increasing the pulse current and duration leads to a higher rate of material removal. Excessive energy input, on the other hand, can result in poor surface quality and increased electrode wear.

Pulse frequency: Higher pulse frequencies improve the rate of material removal. However, if the frequency is too high, the EDM process may become unstable, resulting in reduced machining precision.

Discharge gap: The spacing between the tool and the workpiece has a considerable impact on the EDM process. A narrower gap results in higher precision but slower material removal, whereas a bigger gap results in faster material removal but potentially lower machining accuracy.

Type and pressure of dielectric fluid: The dielectric fluid has an impact on the spark creation and debris cleaning processes. Using deionized water or EDM oils at the proper pressure helps to keep machining conditions stable and results in a superior surface finish.

Duty factor: Duty factor is defined as the ratio of pulse-on time to total cycle time. Higher duty factors result in faster material removal, but they can also cause increased electrode wear and poor surface finish.

Spark gap flushing: Flushing the spark gap effectively ensures stable machining and enhanced surface integrity by removing debris and preventing electrode degradation.

Tool material and shape: The machining accuracy, surface finish, and electrode wear are all influenced by the material and shape of the tool electrode. Electrodes made of copper and graphite are frequently used in EDM.

Workpiece material: The workpiece material's electrical conductivity and hardness have a big impact on the EDM process. Higher discharge energies are typically needed for effective material removal from harder materials.

Dielectric medium temperature: The machining speed and surface quality might be affected by the dielectric fluid's temperature. Improved material removal rates can result from higher temperatures, but too high a temperature can compromise the stability of the machining process.

9.8 TECHNIQUES FOR MODIFYING SURFACES WITH EDM

9.8.1 Conventional surface finishing for EDM

By using EDM to remove material from a workpiece's surface, traditional EDM surface finishing produces a smoother, more polished surface. Using this method, it is possible to achieve precise surface dimensions, remove rough layers, and machine complicated details. Different bioimplant materials, such as titanium, stainless steel, and cobalt-chromium-molybdenum alloys, can be finished using traditional EDM to improve their surface characteristics and biocompatibility.

Traditional EDM surface finishing for bioimplants has the following benefits:

Improved surface roughness: EDM can produce surfaces that are smoother and rougher to a lesser extent, which can improve cell adherence and lessen the chance of bacterial adhesion.

Enhanced biocompatibility: Enhanced smoother surfaces may enhance the biocompatibility of bioimplant materials, lowering the risk of unfavorable host tissue reactions.[23]

Dimensional accuracy: EDM's ability to precisely manage material removal makes it possible to ensure that the bioimplant's dimensions adhere to the required standards.

9.9 SURFACE TEXTURING USING EDM

Surface texturing is a particular use of EDM in which the surface of materials for bioimplants is given a controlled pattern or texture. Simple patterns to intricate microstructures and nanostructures can be found in these textures.[24] Due to its potential to enhance osseointegration, cell adhesion, and wear resistance, surface texturing using EDM has attracted interest in the field of bioimplants.

EDM surface texturing for bioimplants has many benefits:

Enhanced osseointegration: A speedier and more secure osseointegration of the bioimplant is made possible by surface textures that encourage bone cell attachment and development.

Improved wear resistance: Certain textures can limit the bioimplant's contact area with the opposing material, which reduces wear and friction.

Customized surface texturing: Surface texturing can be tailored to meet the unique needs of the bioimplant, giving researchers better control over surface characteristics.

9.10 EDM FOR COATING DEPOSITION

Surfaces of bioimplants can be coated with EDM to increase their biocompatibility, wear resistance, and corrosion resistance. This procedure involves employing EDM to deposit a thin layer of the required coating material onto the surface of the bioimplant.[25] Hydroxyapatite, diamond-like carbon, and biocompatible polymers are typical coating materials.

Benefits of bioimplant EDM coating deposition include:

Improved biocompatibility: Coatings like hydroxyapatite, which resemble the characteristics of real bone, can make bioimplants more biocompatible.

Greater resistance to wear and corrosion: Coatings can give another layer of defense against wear and corrosion, extending the life of the bioimplant.

Flexible coating options: The employment of different coating materials with EDM coating deposition enables the customization of bioimplant surfaces for different medical applications.

9.11 HYBRID TECHNIQUES INCLUDING EDM AND OTHER SURFACE RESURFACING TECHNIQUES

Hybrid processes combine EDM with other surface-treatment techniques to take advantage of the benefits of both approaches. Hybrid methods for bioimplant materials can provide special benefits including better performance and improved surface characteristics.

Hybrid processes for bioimplants examples include:

EDM-assisted laser surface texturing: Using EDM in conjunction with laser surface texturing allows for the creation of intricate patterns and textures on the surfaces of bioimplants, offering special advantages for osseointegration and wear resistance.

EDM-assisted electrochemical machining (ECM): Combining EDM with ECM can increase the control and precision of ECM,[26] enabling better surface finishing and coating deposition on materials used for bioimplants.

EDM-assisted heat treatment: Bioimplant materials can be selectively toughened before heat treatment, resulting in increased wear resistance in particular areas.

EDM-assisted coating modification: EDM can be utilized to alter the coating's surface, resulting in custom surface textures or patterns that improve the performance of bioimplants.

The development of bioimplant technology and the customization of implant surfaces to satisfy the particular needs of various medical applications are both made possible by hybrid processes that combine EDM with other surface treatment techniques. Engineers and medical professionals may develop bioimplants with optimal surface qualities thanks to these methods, which enable the integration of multiple approaches and lead to better patient outcomes.

9.12 EDM SURFACE TREATMENT EFFECTS ON BIOIMPLANT MATERIALS

9.12.1 Improved biocompatibility of surfaces treated with EDM

The surface treatment process known as EDM improves the biocompatibility of bioimplant materials.[27] The procedure produces surfaces that are smoother and have less surface roughness and contamination, which encourages a more beneficial interaction between the bioimplant and adjacent biological tissues. Compared to traditionally machined surfaces, surfaces treated with EDM have demonstrated decreased inflammatory reactions and enhanced cell adhesion, making them more appropriate for implantation.

9.12.2 Influence on the bone-implant interface and osseointegration

Osseointegration, which is essential for the success of orthopedic and dental implants, is positively impacted by EDM surface treatment. EDM surface texturing produces microstructures that increase osseointegration by promoting bone cell adhesion and ingrowth.[28] To improve the bone-implant interface, EDM-assisted coating deposition, such as hydroxyapatite coatings, further encourages bone-implant contact and bone density.

9.12.3 Improved resistance to corrosion

Materials for bioimplants have better corrosion resistance after EDM surface treatment. The procedure changes the surface's chemical makeup, producing a changed layer with improved corrosion resistance.[8] This increases the longevity of the bioimplant by reducing the discharge of metallic ions and the deterioration of the implant's structural integrity.

9.12.4 Influence on wear and friction behavior

Wear and friction behavior are significantly impacted by EDM surface treatment. EDM surface texturing improves wear resistance and decreases friction in bioimplants by lowering wear and frictional coefficients.[28,29] This reduces the production of wear debris and the inflammatory reactions that go along with it, improving implant function over time.

In conclusion, EDM surface treatment influences osseointegration, the bone-implant interface, corrosion resistance, wear and friction behavior, and biocompatibility of bioimplant materials. Due to these results, EDM is a useful surface modification approach for enhancing patient outcomes and bioimplant performance.

9.13 EDM SURFACE MODIFICATION MECHANISMS

9.13.1 Mechanisms for removing material

Through a series of regulated electrical discharges between the tool electrode and the workpiece, EDM removes material. Each discharge produces a high-energy spark that localizes heat at the site of contact, melting and vaporizing the substance. The material is successfully eroded by the small molten particles and debris that are ejected from the workpiece as a result of the tremendous heat and pressure. As the tool electrode advances along the predetermined toolpath, this process is repeated numerous times, progressively shaping the workpiece and producing the necessary surface alterations.

9.13.2 Recrystallization and surface melting

Localized electrical discharges produced by EDM cause the workpiece material's surface to melt. The material melts and forms a thin coating of molten material on the surface due to the high temperatures attained during the spark.[30] The molten substance rapidly cools and hardens after the spark is extinguished, causing recrystallization.

EDM surface melting and recrystallization might change the material's microstructure. This could affect the bioimplant material's performance and behavior by influencing parameters like hardness, grain size, and residual stresses.

9.13.3 Surface texture enhancement

Through carefully controlled tool movement and material removal during EDM, surface texture development is accomplished. Different surface textures can be created by adjusting the EDM process's pulse frequency, pulse duration, and discharge gap. When EDM is used to surface texturize bioimplants in order to increase osseointegration and improve cell adhesion, surface texture creation is very important.[31]

The topography and roughness of the bioimplant surface can be influenced by the surface texture created by EDM, which in turn impacts how the bioimplant interacts with the surrounding biological environment.

9.13.4 Phase transformation and residual stresses

EDM involves localized high temperatures that can cause phase changes in some materials. The crystal structure and mechanical characteristics of the material may change as a result of these phase transitions.

Additionally, residual stresses in the material may be introduced by the nonuniform heating and cooling that occurs during EDM. The mechanical behavior and fatigue resistance of the bioimplant material can be influenced by residual stresses, which can take both compressive and tensile forms.[32]

To optimize the EDM process and modify the surface properties of bioimplant materials to meet particular medical needs, it is crucial to comprehend the mechanisms of EDM surface modification, including material removal, surface melting and recrystallization, surface texture formation, phase transformation, and residual stresses. Through the use of these mechanisms, scientists and engineers can create bioimplants with increased biocompatibility, osseointegration, corrosion resistance, and wear behavior, which enhances patient outcomes and increases the success of the implants over the long run.

9.14 CASE STUDIES

9.14.1 Titanium alloys treated with EDM for dental implants

Dental implants made of titanium alloy that had undergone EDM treatment had much better surface properties than implants that had undergone conventional machining. Surfaces became smoother as a result of the surface roughness being reduced. In vitro tests on the EDM-treated surfaces exhibited increased cell adhesion, proliferation, and differentiation, indicating improved biocompatibility.[33] Enhanced bone-to-implant contact and greater bone density around the EDM-treated implants were observed in the in vivo animal experiments, indicating improved osseointegration. Overall, the study showed that EDM surface treatment enhanced titanium alloy dental implants' biocompatibility and osseointegration, making them more appropriate for long-term clinical use.

9.14.2 Orthopedic implant surface texturing using EDM

The orthopedic implants' wear resistance was dramatically increased by the EDM surface texturing. When compared to implants that had been normally machined, the textured surfaces demonstrated lower wear rates and frictional coefficients. EDM's micro-dimples and grooves served as lubricant reservoirs, lowering friction and reducing direct contact during joint movement.[34] By minimizing wear and friction, wear debris formation also decreased, lowering the risk of inflammatory reactions and possible implant failure. The results of this study showed that EDM surface texturing could significantly improve the wear behavior and friction characteristics of orthopedic implants made of the cobalt-chromium-molybdenum alloy, making them more robust and appropriate for long-term joint replacement applications.

9.14.3 EDM coating deposition for better biocompatibility

The hydroxyapatite (HA) coatings created by EDM showed high biocompatibility. Studies done in vitro revealed that coated surfaces of titanium implants had better cell adhesion, proliferation, and osteogenic differentiation than uncoated surfaces. The HA-coated implants had greater bone-to-implant contact and more bone density around them, which suggests superior osseointegration, according to the in vivo animal experiments.[35] The implants with HA coating demonstrated greater integration with the host bone, facilitating a quicker and more reliable fixation. This case study showed that coating deposition with EDM assistance might significantly improve the biocompatibility and osseointegration of titanium bioimplants, presenting a promising strategy to raise implant success rates and patient outcomes.

9.14.4 Nitinol stent surface modification by EDM

The nitinol stents' smoother surfaces and fewer surface flaws were produced using ED surface modification. The distinctive superelastic and shape memory features of nitinol were preserved because the method did not materially change the bulk material's properties. In vitro biocompatibility experiments on the EDM-treated stents revealed increased cell adhesion and proliferation, demonstrating improved biocompatibility.[36] According to the study, EDM surface modification could successfully improve the mechanical and biocompatibility of nitinol stents, potentially providing advantages for cardiovascular treatments.

9.14.5 Surface texturing of dental implant abutments using EDM

The dental implant abutments' well-defined microstructures were produced by EDM surface texturing. For the purpose of promoting fibroblast cell adhesion and proliferation in vitro, the surface roughness and topography were tuned. In comparison to typical smooth abutments, the clinical trial showed better soft tissue adhesion and stability around the EDM-textured abutment.[37] The textured abutments improved patient comfort and aesthetics, demonstrating successful clinical outcomes. This study showed how EDM surface texturing could improve the stability and soft tissue integration of dental implant restorations.

9.15 BIOLOGICAL ASSESSMENT AND SAFETY CONSIDERATIONS

9.15.1 Evaluations of in vitro biocompatibility

Evaluations of in vitro biocompatibility are essential for determining how biological systems at the cellular level interact with bioimplants that have undergone EDM treatment. In order to measure cell adhesion, proliferation, viability, and cellular responses to the treated surfaces, these tests are commonly carried out utilizing cell culture experiments. These studies involve the use of a variety of cell types pertinent to the particular bioimplant application, such as endothelial cells for cardiovascular devices or osteoblasts for orthopedic implants.[38] These evaluations aid in figuring out whether the EDM-treated surfaces encourage beneficial biological responses, such as increased cell adhesion and proliferation, decreased cytotoxicity, and improved tissue integration. Additionally, they offer important information about any negative effects or inflammatory reactions, assisting in the optimization of EDM parameters and surface alterations for improved biocompatibility.

9.15.2 In vivo animal studies

For assessing the security and effectiveness of EDM-treated bioimplants in living creatures, in vivo animal investigations are crucial. The biocompatibility, tissue response, and long-term effectiveness of the implants are evaluated using animal models, such as mice or larger animals. These investigations involve implanting the EDM-treated bioimplants into animal subjects and seeing how they interact with the tissues and organs around them over a predetermined time. To determine the effectiveness and safety of the implants, factors such as tissue integration, osseointegration, inflammatory response, and implant degradation are evaluated.[2] Between in vitro evaluations and human therapeutic trials, animal research provides useful preclinical data.

9.15.3 Clinical assessments of bioimplants treated with EDM

Clinical evaluations include setting EDM-treated bioimplants through rigorous clinical trials on real patients. These studies are carried out to assess the implant's performance, effectiveness, and safety in actual patient populations. Clinical analyses examine a number of variables, such as implant stability, biocompatibility, tissue fusion, patient comfort, and overall therapeutic success. The effects and consequences of the EDM-treated bioimplants are continuously evaluated in the patients.[39] To determine the longevity and success rates of the implants, long-term follow-up is carried out. Positive clinical evaluation outcomes demonstrate the safety and efficacy of the EDM-treated bioimplants, resulting in regulatory approval and broad use in clinical practice.

9.15.4 Safety factors and regulatory concerns

When adopting EDM surface treatment for bioimplants, safety must come first. The integrity, mechanical capabilities, or biocompatibility of the material should not be jeopardized by surface alterations. When too much material is removed during EDM, the implant may become weaker and lose structural integrity. The approval and commercialization of bioimplants are regulated by agencies like the European Medicines Agency (EMA) in the European Union and the U.S. Food and Drug Administration (FDA) in the United States.[40] EDM-treated bioimplants must adhere to applicable medical device laws and go through extensive testing to prove their safety and effectiveness. For bioimplants that have had EDM treatment, manufacturers are required to carry out extensive risk analyses, biocompatibility testing, and sterilization validations. Additionally, they must offer thorough documentation of the EDM procedure, its variables, and the surface alterations made. In order to guarantee consistent quality, safety, and performance of EDM-treated bioimplants, adherence to Good Manufacturing Practices (GMP) and ISO standards for medical devices is essential. In conclusion, the development and marketing of EDM-treated bioimplants require key steps in biological evaluation, safety evaluation, and regulatory considerations. To ensure the effectiveness of the implants in medical applications and enhance patient outcomes, a thorough understanding of the biocompatibility, tissue response, and long-term performance of the implants is necessary.

9.16 FUTURE PERSPECTIVES AND CHALLENGES

9.16.1 Potential improvements in EDM technology for use with bioimplants

Several prospective developments in EDM technology show promise for bioimplant applications:

Enhanced precision and surface quality: Improvements in the design and control of EDM machines may result in improved precision and surface quality, enabling the production of bioimplants with even finer surface textures and complex geometries.

Multiaxis EDM: Multiaxis EDM machines may be used to create patient-specific implants with incredibly detailed geometries and specialized surface modifications, enhancing implant functionality and integrating implants even more.

Advanced material compatibility: Expanding EDM's compatibility with a wider variety of bioimplant materials, such as bioresorbable polymers, ceramics, and composites, may be the focus of research efforts.

In-process monitoring and feedback systems: By incorporating real-time monitoring and feedback systems during EDM, it may be possible to reduce manufacturing variability by optimizing process parameters and ensuring uniform surface alterations.

9.16.2 EDM integration with additive manufacturing processes

The production of bioimplants has a lot of potential when EDM is combined with additive manufacturing techniques like 3D printing or selective laser melting:

Customized implants: By combining EDM with additive manufacturing, it is possible to create bioimplants that are specifically designed for each patient, addressing their specific anatomical needs and increasing implant success rates.

Gradient materials: By combining EDM with additive manufacturing, gradient materials could be used to fabricate bioimplants, allowing for the customization of both surface and mechanical properties to enhance implant effectiveness.

Hybrid implants: Hybrid methods that combine additively built implant structures with EDM-treated surfaces may offer a combination of superior mechanical characteristics and improved biocompatibility.

9.17 CONSIDERATIONS FOR SUSTAINABILITY AND THE ENVIRONMENT

As with any manufacturing process, environmental and sustainability considerations are important for EDM's continued use in bioimplant applications:

Material recycling: Recycling and reusing waste materials produced by EDM, such as removed metal particles and dielectric fluids, can lessen their negative effects on the environment and increase their sustainability.

Ecofriendly dielectric fluids: The ecological impact of EDM procedures can be reduced through research into biodegradable and environmentally friendly dielectric fluids.

Green manufacturing techniques: Green manufacturing techniques, including the use of energy-efficient equipment and waste minimization techniques, can support environmental sustainability in EDM facilities.

9.18 FUTURE RESEARCH DIRECTIONS AND CHALLENGES

While EDM has a lot of potential for changing the surface of bioimplants, there are still a number of problems and areas for future research:

Long-term implant performance: To evaluate the endurance and functionality of EDM-treated bioimplants over long periods of time, long-term research on the performance and biocompatibility of these implants is crucial.

Standardization and regulation: Establishing regulatory requirements and testing methodologies that are specific to EDM-treated bioimplants will guarantee their consistency in terms of quality and safety in clinical applications.

Understanding material–process interactions: Additional investigation is required to fully grasp how EDM surface treatment impacts different bioimplant materials, their mechanical characteristics, and long-term behavior.

Multidisciplinary collaboration: To advance EDM technology for bioimplant applications, multidisciplinary cooperation among material scientists, engineers, biologists, physicians, and regulatory specialists is essential.

Cost-effectiveness: Ongoing efforts to improve throughput, lower production costs, and optimize the EDM process will increase the economic viability of bioimplants that have undergone EDM treatment.

In conclusion, EDM has tremendous promise for the future in bioimplant applications, with ongoing research, integration with additive manufacturing, environmental concerns, and technological developments all helping to create safer and more effective bioimplant solutions. Transformative developments in the field of medical implants will be made possible by addressing remaining issues and promoting multidisciplinary cooperation.

9.19 CONCLUSIONS

9.19.1 Summary of key findings

Several significant discoveries have been made as a result of the investigation into EDM as a surface modification technique for bioimplant materials:

Improved biocompatibility: Bioimplants that have undergone EDM treatment have shown enhanced biocompatibility, encouraging cell adhesion, proliferation, and differentiation on the treated surfaces. For implant integration and tissue regeneration to be effective, this improved biocompatibility is essential.

Osseointegration and bone-implant interface: EDM surface treatment has the ability to improve osseointegration, which results in improved bone-to-implant contact and higher bone density around the treated implants. This is crucial for the stability and functionality of orthopedic and dental implants over the long term.

Improvements in corrosion resistance: Bioimplants that have had EDM treatment have shown improved corrosion resistance, which is essential for preventing implant deterioration and reducing unfavorable effects on surrounding tissues.

Wear and friction behavior: In particular in load-bearing applications, EDM surface texturing and coating deposition have proved successful in enhancing wear resistance and lowering frictional coefficients of bioimplants.

Personalization and customization: EDM makes it possible to design highly personalized implant surfaces that are patient-specific, allowing for the best possible clinical results by adjusting them to each person's unique anatomical needs.

Antimicrobial properties: EDM surface modification has the potential to produce antimicrobial surfaces that prevent bacterial colonization and adherence, lowering the risk of implant-associated illnesses.

9.19.2 Promising future prospects

The potential for EDM in bioimplant applications is quite positive:

Technological developments: EDM technology will continue to evolve thanks to improvements in precision, multiaxis capabilities, and real-time monitoring systems. This will further improve surface modifications and make it possible to create complicated implant designs.

Integration with additive manufacturing: Combining EDM with additive manufacturing techniques will enable the creation of hybrid implant structures, gradient materials, and personalized implants with enhanced mechanical and biocompatibility properties.

Environmental sustainability and ecofriendly practices: By emphasizing recycling and using ecofriendly dielectric fluids, EDM surface treatment will be in line with environmentally friendly manufacturing techniques.

Clinical translation: More clinical assessments and regulatory clearances will make it easier to translate bioimplants treated with EDM from the lab to clinical practice, which will help patients by enhancing their chances of a successful course of therapy.

9.20 FURTHER RESEARCH SUGGESTIONS

It is advised that more study be done in the following fields to fully realize the promise of EDM in bioimplant applications:

Long-term performance studies: Comprehensive long-term studies should be conducted to evaluate the functionality, longevity, and safety of EDM-treated bioimplants in vivo.

Material-process relationships: To maximize surface changes and reduce unfavorable effects, further understanding of the relationships between EDM process parameters and diverse bioimplant materials is needed.

Standardization and regulation: To guarantee consistency in quality and safety in clinical applications, develop standardized testing methodologies and regulatory requirements unique to EDM-treated bioimplants.

Multidisciplinary collaboration: Collaboration across disciplines is encouraged in order to overcome obstacles and develop the science of EDM-treated bioimplants. This group of professionals includes materials scientists, engineers, biologists, physicians, and regulatory specialists.

Cost-effectiveness: To increase the economic feasibility of EDM-treated bioimplants and increase patient accessibility, concentrate on cost-effective production techniques.

The biocompatibility, osseointegration, corrosion resistance, wear behavior, and customization of bioimplants all stand to benefit greatly from EDM surface treatment. Successful clinical use of bioimplants treated with EDM will greatly enhance patient outcomes and progress medical treatments.

Unlocking the full potential of EDM in the field of bioimplants requires embracing sustainability, encouraging collaboration, and spending money on more research.

REFERENCES

1. Sumayli, A. (2021). Recent trends on bioimplant materials: A review. *Materials Today: Proceedings*, 46, 2726–2731. doi:10.1016/j.matpr.2021.02.395
2. Kang, C.-W., & Fang, F.-Z. (2018). State of the art of bioimplants manufacturing: Part II. *Advances in Manufacturing*, 6(2), 137–154. doi:10.1007/s40436-018-0218-9
3. Mohd Abbas, N., Solomon, D. G., & Fuad Bahari, M. (2007). A review on current research trends in electrical discharge machining (EDM). *International Journal of Machine Tools and Manufacture*, 47(7–8), 1214–1228. doi:10.1016/j.ijmachtools.2006.08.026
4. Mahajan, A., Sidhu, S. S., & Ablyaz, T. (2019). EDM surface treatment: An enhanced biocompatible interface. In: Bains, P., Sidhu, S., Bahraminasab, M., Prakash, C. (eds) *Biomaterials in Orthopaedics and Bone Regeneration. Materials Horizons: From Nature to Nanomaterials*. Springer, Singapore. https://doi.org/10.1007/978-981-13-9977-0_3
5. Aliyu, A. A., Abdul-Rani, A. M., Ginta, T. L., Prakash, C., Axinte, E., Razak, M. A., & Ali, S. (2017). A review of additive mixed-electric discharge machining: Current status and future perspectives for surface modification of biomedical implants. *Advances in Materials Science and Engineering*, 2017, 1–23. doi:10.1155/2017/8723239
6. Pujiyulianto, E., & Suyitno. (2021). Effect of pulse current in manufacturing of cardiovascular stent using EDM die-sinking. *The International Journal of Advanced Manufacturing Technology*, 112(11–12), 3031–3039. doi:10.1007/s00170-020-06484-3
7. Grimberg, R., Udpa, L., Savin, A., Chan, S. C., Steigmann, R., & Udpa, S. S. (2009). Noninvasive evaluation of Bjork-Shiley convexo-concave prosthetic heart valves. *NDT & E International*, 42(5), 421–429. doi:10.1016/j.ndteint.2009.01.013
8. Kamachimudali, U., Sridhar, T. M., & Raj, B. (2003). Corrosion of bio implants. *Sadhana*, 28, 601–637. doi:10.1007/BF02706450
9. Ralls, A., Kumar, P., Misra, M., & Menezes, P. L. (2019). Material design and surface engineering for bio-implants. *JOM*. doi:10.1007/s11837-019-03687-2
10. Shen, G., Zhang, J., Melentiev, R., & Fang, F. (2021). Study on tribological performance of groove-textured bioimplants. *Journal of the Mechanical Behavior of Biomedical Materials*, 119, 104514. doi:10.1016/j.jmbbm.2021.104514
11. Madhan Kumar, A., Nagarajan, S., Ramakrishna, S., Sudhagar, P., Kang, Y. S., Kim, H., … Rajendran, N. (2014). Electrochemical and in vitro bioactivity of polypyrrole/ceramic nanocomposite coatings on 316L SS bioimplants. *Materials Science and Engineering: C*, 43, 76–85. doi:10.1016/j.msec.2014.07.005
12. Matos, G. R. M. (2020). Surface roughness of dental implant and osseointegration. *Journal of Maxillofacial and Oral Surgery*, 20(1), 1–4. doi:10.1007/s12663-020-01437-5

13. Parau, A. C., Juravlea, G. A., Raczkowska, J., Vitelaru, C., Dinu, M., Awsiuk, K., Vranceanu, D. M., Ungureanu, E., Cotrut, C. M., & Vladescu, A. (2023). Comparison of 316L and Ti6Al4V biomaterial coated by ZrCu-based thin films metallic glasses: Structure, morphology, wettability, protein adsorption, corrosion resistance, biomineralization. *Applied Surface Science*, 612, 155800. doi:10.1016/j.apsusc.2022.155800
14. Malaisrie, S. C., Malekzadeh, S., & Biedlingmaier, J. F. (1998). In vivo analysis of bacterial biofilm formation on facial plastic bioimplants. *The Laryngoscope*, 108(11), 1733–1738. doi:10.1097/00005537-199811000-00026
15. Asri, R. I. M., Harun, W. S. W., Samykano, M., Lah, N. A. C., Ghani, S. A. C., Tarlochan, F., & Raza, M. R. (2017). Corrosion and surface modification on biocompatible metals: A review. *Materials Science and Engineering: C*, 77, 1261–1274. doi:10.1016/j.msec.2017.04.102
16. Farooq, M. U., Bhatti, H. A., Asad, M., et al. (2022). Surface generation on titanium alloy through powder-mixed electric discharge machining with the focus on bioimplant applications. *International Journal of Advanced Manufacturing Technology*, 122, 1395–1411. doi:10.1007/s00170-022-09927-1
17. Ho, K., & Newman, S. (2003). State of the art electrical discharge machining (EDM). *International Journal of Machine Tools and Manufacture*, 43(13), 1287–1300. doi:10.1016/s0890-6955(03)00162-7
18. Ho, K., Newman, S., Rahimifard, S., & Allen, R. (2004). State of the art in wire electrical discharge machining (WEDM). *International Journal of Machine Tools and Manufacture*, 44(12–13), 1247–1259. doi:10.1016/j.ijmachtools.2004.04.017
19. Khajuria, A., Akhtar, M., Pandey, M. K., Singh, M. P., Raina, A., Bedi, R., & Singh, B. (2019). Influence of ceramic Al2O3 particulates on performance measures and surface characteristics during sinker EDM of stir cast AMMCs. *World Journal of Engineering*, 16(4), 526–538. doi:10.1108/wje-01-2019-0015
20. Li, G., Natsu, W., & Yu, Z. (2019). Study on quantitative estimation of bubble behavior in micro hole drilling with EDM. *International Journal of Machine Tools and Manufacture*, 103437. doi:10.1016/j.ijmachtools.2019.103437
21. Ye, S. L., Su, J., Jia, Z. X., & Hua, R. (2004). Study on the mechanism of high speed small hole drilling by EDM. *Materials Science Forum*, 471–472, 302–306. doi:10.4028/www.scientific.net/msf.471-472.302
22. Jabbaripour, B., Sadeghi, M. H., Faridvand, S., & Shabgard, M. R. (2012). Investigating the effects of EDM parameters on Surface Integrity, MRR and TWR in machining of Ti–6Al–4V. *Machining Science and Technology*, 16(3), 419–444. doi:10.1080/10910344.2012.698971
23. Prakash, C., Singh, S., Pabla, B. S., & Uddin, M. S. (2018). Synthesis, characterization, corrosion and bioactivity investigation of nano-HA coating deposited on biodegradable Mg-Zn-Mn alloy. *Surface and Coatings Technology*, 346, 9–18. doi:10.1016/j.surfcoat.2018.04.035
24. Guu, Y. H., & Hou, M. T.-K. (2007). Effect of machining parameters on surface textures in EDM of Fe-Mn-Al alloy. *Materials Science and Engineering: A*, 466(1–2), 61–67. doi:10.1016/j.msea.2007.02.035

25. Subramanian, B., Muraleedharan, C. V., Ananthakumar, R., & Jayachandran, M. (2011). A comparative study of titanium nitride (TiN), titanium oxy nitride (TiON) and titanium aluminum nitride (TiAlN), as surface coatings for bio implants. *Surface and Coatings Technology*, 205(21–22), 5014–5020. doi:10.1016/j.surfcoat.2011.05.004
26. Hui, C. (2019). Experimental study of electrochemical micro-milling of silicon. *IOP Conference Series: Earth and Environmental Science*, 267, 062058. doi:10.1088/1755-1315/267/6/062058
27. Lee, B. E. J., Ho, S., Mestres, G., Karlsson Ott, M., Koshy, P., & Grandfield, K. (2016). Dual-topography electrical discharge machining of titanium to improve biocompatibility. *Surface and Coatings Technology*, 296, 149–156. doi:10.1016/j.surfcoat.2016.04.024
28. Prakash, C., Kansal, H. K., Pabla, B. S., & Puri, S. (2017). On the influence of nanoporous layer fabricated by PMEDM on β-Ti implant: Biological and computational evaluation of bone-implant interface. *Materials Today: Proceedings*, 4(2), 2298–2307. doi:10.1016/j.matpr.2017.02.078
29. Shen, G., Fang, F., & Kang, C. (2018). Tribological performance of bioimplants: A comprehensive review. *Nanotechnology and Precision Engineering*, 1, 107–122. doi:10.13494/j.npe.20180003
30. Hashmi, A. W., Mali, H. S., Meena, A., Saxena, K. K., Ahmad, S., Agrawal, M. K., Sagbas, B., Valerga Puerta, A. P., & Ijaz Khan, M. (2023). A comprehensive review on surface post-treatments for freeform surfaces of bio-implants. *Journal of Materials Research and Technology*, 23, 4866–4908. doi:10.1016/j.jmrt.2023.02.007
31. Mahajan, A., & Sidhu, S. S. (2017). Surface modification of metallic biomaterials for enhanced functionality: A review. *Materials Technology*, 33(2), 93–105. doi:10.1080/10667857.2017.1377971
32. Al-Amin, M., Abdul Rani, A. M., Abdu Aliyu, A. A., Abdul Razak, M. A., Hastuty, S., & Bryant, M. G. (2020). Powder mixed-EDM for potential biomedical applications: A critical review. *Materials and Manufacturing Processes*, 1–23. doi:10.1080/10426914.2020.1779939
33. Singh, V., Katyal, P., Kumar, K., & Kumar, R. (2022). Surface integrity and biological response of Ti-Alloy implants after surface modification. *Materials Today: Proceedings*, 56(5), 2451–2468. doi:10.1016/j.matpr.2021.08.235
34. Davis, R., Singh, A., Jackson, M.J., et al. (2022). A comprehensive review on metallic implant biomaterials and their subtractive manufacturing. *International Journal of Advanced Manufacturing Technology*, 120, 1473–1530. doi:10.1007/s00170-022-08770-8
35. Al-Amin, M., Abdul Rani, A. M., Abdu Aliyu, A. A., Bryant, M. G., Danish, M., & Ahmad, A. (2020). Bio-ceramic coatings adhesion and roughness of biomaterials through PM-EDM: A comprehensive review. *Materials and Manufacturing Processes*, 1–24. doi:10.1080/10426914.2020.1772483
36. Shabalovskaya, S., Anderegg, J., & Van Humbeeck, J. (2008). Critical overview of Nitinol surfaces and their modifications for medical applications. *Acta Biomaterialia*, 4(3), 447–467. doi:10.1016/j.actbio.2008.01.013
37. Mehl, A. C. S. B., Benegra, M., & Pintaude, G. (2021). Surface characterization of the seating platform of titanium implant processed with different textures. *Journal of the Brazilian Society of Mechanical Sciences and Engineering*, 43, 130. doi:10.1007/s40430-021-02853-5

38. Mahajan, A., & Sidhu, S. S. (2018). Enhancing biocompatibility of Co-Cr alloy implants via electrical discharge process. *Materials Technology*, 33(8), 524–531. doi:10.1080/10667857.2018.1475144
39. Mahajan, A., & Sidhu, S. S. (2019). In vitro corrosion and hemocompatibility evaluation of electrical discharge treated cobalt–chromium implant. *Journal of Materials Research*, 34(8), 1363–1370. doi:10.1557/jmr.2019.73
40. Zhou, L., GJVM, V. O., Malda, J., Stoddart, M. J., Lai, Y., Richards, R. G., … Qin, L. (2020). Innovative tissue-engineered strategies for osteochondral defect repair and regeneration: Current progress and challenges. *Advanced Healthcare Materials*, 9(23), 2001008. doi:10.1002/adhm.202001008

Index